FUSARIUM

Laboratory Guide to the Identification of the

Major Species

by

C. Booth

Assistant Director

Commonwealth Mycological Institute, Ferry Lane, Kew, Surrey

Commonwealth Mycological Institute

Kew, Surrey, England

1977

FUSARIUM

Laboratory Guide to the Identification of the

Major Species

by

C. Booth

Assistant Director

Commonwealth Mycological Institute, Ferry Lane, Kew, Surrey

Commonwealth Mycological Institute

Kew, Surrey, England

1977

First published March 1977 by the Commonwealth Mycological Institute
under the authority of the
Executive Council, Commonwealth Agricultural Bureaux,
Farnham Royal, Slough, England

ISBN 0 85198 383 9

This and other publications of the
Commonwealth Agricultural Bureaux
can be obtained through any major bookseller or
direct from
Central Sales Branch, Commonwealth Agricultural Bureaux,
Farnham Royal, Slough SL2 3BN, England

CONTENTS

INTRODUCTION

This laboratory guide is an attempt to illustrate the diagnostic characters which are used to separate the most frequently occurring *Fusarium* species, and to provide a key for their identification. Those species occurring on insects or as parasites of Sphaeriaceous fungi are not included.

For further information on *Fusarium* species, including details of synonymy, hosts and geographical distribution, reference can be made to the book 'The Genus Fusarium' by C. Booth (Commonwealth Mycological Institute, 1971).

The characters used in the key are those which can be observed after 7-10 days of growth in culture. Perithecial states are mentioned where known, but they seldom develop on normal culture plates, except in *F. xylarioides*, in which perithecial initials form readily, or in certain homothallic strains of *F. solani*.

SPECIES KEY

1. Cultures slow growing : growth rate below 2 cm   27
1. Growth rate above 2 cm, usually 4-8 cm   2
2. Conidia variable, no clear distinction between micro and macroconidia
     14
2. Conida uniform or may be distinction between micro and macroconidia
     3
3. Microconidia present in abundance, macroconidia present or absent
     4
3. Microconidia sparse or absent, macroconidia usually present
     16
4. Microconidia formed in chains.   5
4. Microconidia not formed in chains.   6
5. Cultures creamy, pale beige to deep violet ; macroconidia scattered often sparse
or absent*   *F. moniliforme* (3)
5. Cultures crimson-rose ; macroconidia formed in creamy-yellow pustules
     F. decemcellulare (6)
6. Microconidia formed from simple phialides, borne laterally on hyphae or on
conidiophores.   7
6. Microconidia formed as blastospores or from polyphialides
     11
7. Culture pigmentation beige, blue, violet or white   8
7. Culture pigmentation red, vinaceous   10
8. Microconidia globose, pale or non-pigmented strains *F. poae* (7)
8. Microconidia oval, pyriform or cylindrical   9
9. Microconidiophores well developed with long phialides, often elaborately
branched after a few days.   *F. solani* (2)
9. Microconidia formed from lateral phialides, microconidiophores poorly developed often absent or at most a foot cell with 1-3 apical phialides
     F. oxysporum (1)
10. Microconidia globose   *F. poae* (7)
10. Microconidia ovoid to pyriform   *F. tricinctum* (8)
11. Microconidia formed from polyblastic conidiogenous cells
     12
11. Microconidia formed, at least in part, from polyphialides
     13
12. Microconidia fusiform to clavate, 8-12 × 2.5-4μ *F. fusarioides* (15)
12. Microconidia obovate to clavate 6-20 × 2.5-4μ *F. sporotrichioides* (16)
13. Microconidia clavate   *F. moniliforme* v.
     subglutinans (4)
13. Both clavate and globose to oval microconidia present *F. moniliforme* v.
     anthophilum (5)
14. Conidia 1-3 septate, more or less evenly curved *F. nivale* (25)
14. Conidia 1-5 septate with uncinate or hooked apical cell
     15
15. Cultures sulphureous, pink becoming salmon-orange, occasional strains with
violet pigment. No perithecial initials.   *F. udum* (13)
15. Cultures colourless, pale peach with adpressed mycelium, violet pigmentation
formed with production of perithecial initials *F. xylarioides* (14)
16. First formed conidia produced from polyblastic conidiogenous cells
     17
16. Conidia formed from simple phialides   18
17. Primary conidia with wedge-shaped foot cells ; cultures from below beige brown
becoming dark brown   *F. semitectum* (17)
17. Primary conidia with pointed or fusoid foot cell ; cultures from below red to
reddish brown.   *F. avenaceum* (18)
18. Macroconidia with uncinate or hooked apical cell *F. lateritium* (11)
18. Macroconidia with curved, beaked or elongated apical cell
     19

***Pigmentation throughout based on growth on PSA agar, pH 6.5-7.**

19. Cultures from below pink, beige, salmon, sulphureous to brown
    20

19. Cultures from below carmine red to dark reddish purple
    22

20. Cultures from below sulphureous ; conidia 3-4 septate, 23-32 × 3.5-4μ
    F. sulphureum (21)

20. Cultures from below colourless, pink or beige becoming pale to dark brown ; conidia up to 60μ long   21

21. Chlamydospores in knots or chains ; cultures beige becoming pale to dark brown
    F. equiseti (9)

21. Chlamydospores extremely sparse usually absent ; cultures pale pink or peach
    F. heterosporum (20)

22. Apex of macroconidia elongated, narrowing evenly to a point or with extreme elongation of apical cell   23

22. Apical cell of macroconidia strongly curved to ventral side to form a beak
    25

23. Chlamydospores formed in knots or chains   *F. acuminatum* (10)
23. Chlamydospores sparse or absent   24
24. Conidia 30-60 × 3.5-5μ ; chlamydospores may or may not be present
    F. graminearum (19)

24. Conidia 40-80 × 2.5-4μ ; chlamydospores always absent
    F. avenaceum (18)

25. Conidia not more than 5μ wide   *F. sambucinum* (22)
25. Conidia between 5 and 7μ wide   26
26. Conidia 30-50 × 5-7μ *F. culmorum* (24)
26. Conidia 25-35 × 5-6μ *F. sambucinum* v. coeruleum (23)

27. Species occurring on, or isolated from, Sphaeriaceous fungi. (These species are not included in this key).

27. Isolates not originating from Sphaeriaceous fungi   28
28. Conidia over 30μ long   29
28. Conidia 30μ or less in length   30
29. Macroconidia 30-70 × 3.5-4μ *F. merismoides* (29)
29. Macroconidia variable but 3-5 septate are 40-65 × 3-4μ
    F. aquaeductuum (Aggr.) (28)

30. Conidia strongly curved and pointed at the apex, chlamydospores often present
    F. dimerum (26)

30. Conidia gently curved with rounded apex : chlamydospores absent
    31

31. Conidia 15-30 × 3-5μ *F. nivale* (25)
31. Conidia 12-16 × 3-4μ *F. tabacinum* (27)

METHODS AND MEDIA

Isolations from Soil

Fusarium species are of frequent occurrence in most soils and because of their competitive ability can be easily isolated in the presence of Phycomycetes, dry-spored moulds, Actinomycetes or bacteria. In order to do this a weak medium such as tap water agar (media 4 below) or a specially selected medium such as 5 below should be used. The surface of the agar should be allowed to dry in a cool dark place so that any water from a dilute soil or spore suspension is absorbed by the dried agar ; this prevents the rapid spread of bacteria, actinomycetes or phycomycetous fungi into a developing *Fusarium* colony.

Subcultures should be taken from the edge of a developing colony as soon as this can be observed under a dissecting microscope (see Nash & Snyder, 1962).

Isolations from Plant Material

Fusarium sporòdochia found on woody stems or other plant tissue should be moistened and a dilute spore suspension made from the mass of spores.

Fusarium species causing wilt penetrate the vascular tissue in the stem or roots. These are best isolated by surface sterilising small segments or sections of the material preferably with Chlorox (sodium hypochlorite) although mercuric chloride, alcohol, hydrogen peroxide or silver nitrate can also be used (see chart). After a final thorough washing in distilled water the ends of the tissue are sliced off with a sterile scalpel and the centre sections placed on a PDA or PSA (see media 1 and 2) plate. After 24-36 h at 20-25°C (14-18° for suspected *F. nivale*) *Fusarium* mycelium should have grown out sufficiently from the plant material to be re-isolated on a suitable agar plate. For identification purposes PSA (see media 1) has been found eminently suitable but not essential. PDA or OA are also satisfactory.

Single Spore Isolation

There is no evidence for the assumption that *Fusarium* mycelium growing out of plant tissue or from soil particles will necessarily belong to a single species and much confusion has been caused by trying to identify mixed cultures. To avoid this and to ensure isolates are pure, working with single spore isolations is strongly recommended. In fact a series of isolates grown from single spores demonstrate species characteristics and uniform growth much more clearly than mass transfer of inoculum.

The simplest and most economical way of obtaining a series of single spore cultures is as follows :—

A drop of sterile water is placed on a sterile slide under the dissecting microscope. An accumulation of spores is obtained on the wet tip of a needle either from a sporodochium or from aerial mycelium. The tip of the needle is then introduced into the drop of water on the slide and the spores can be observed to flow from the tip of the needle into the drop of water. When the suspension is adequate the point of the needle with any remaining spores is withdrawn. Experience of the correct dilution can easily be acquired ; it is approximately the point when the individual spores are distinguishable in the water drop and not obscured by overlapping. (If the concentration is too dense difficulties will be experienced in subsequently removing single spores from the plate.) The spore suspension on the slide is then picked up by a sterile loop and streaked across a clear agar plate, the position of the streaks being marked by crayon lines on the bottom of the petri-dish. The plate is then incubated for 12-16 h at 25°C. By following the lines under the low power of a compound microscope germination of the conidia can be observed and those which are suitably clearly positioned can be removed on a small 1 mm square of the agar by a fine knife or flattened tip of a needle and transferred to a petri-dish or test tube.

Stimulation of Sporulation

Subsequent growth of the isolated spores will form a colony suitable for identification if incubated at 22-25°C in the light, for 7-10 days. In single spore isolates abundant sporulation usually occurs after this time. This is not necessarily true with old cultures that have become stale by transmission through the post or because of prolonged growth in artificial culture. Spore production can usually be stimulated in staled cultures by the use of one of the following methods :—

1. Single spore cultures obtained from some of the few, often abnormal spores usually present in staled cultures.
2. Scraping off the aerial mycelium, washing the surface of the agar with several changes of sterile distilled water and re-incubation of plate.
3. Placing 1 cm squares of agar cut from the culture into sterile water in a sterile petri-dish.
4. Re-subculture and incubate at 14-18°C.
5. Place 3-4 day old cultures under a bank of two daylight or 'true light' fluorescent tubes and one near-ultraviolet (black light) tube on a 12 hr on/off cycle.

Strains of *Fusarium graminearum* which do not sporulate readily often respond to methods 2 or 3 (see also media 7). *F. heterosporum* sporulates abundantly under method 5. Non-sporulating strains of *F. nivale* and *F. aquaeductuum* respond to method 4.

Sporogenous Cells and Slide Cultures

The method of spore production and the nature of the sporogenous cell is a critical character in the identification of *Fusarium* species. The best method of observation of these and especially if one is unfamiliar with their appearance is to grow them in slide culture (Riddell, 1950). Such cultures are simple to prepare and easily converted into semi-permanent reference slides.

Slide Cultures

Pour a plate of suitable agar about 3 mm thick and for each slide culture cut out a 6 mm square. This is placed in the centre of a sterile slide and inoculated on each of the vertical faces with a minute quantity of inoculum on the point of a needle. A sterile cover slip is then placed on top of the agar. The slide culture is then placed in a damp chamber and incubated until the mycelium which develops on each face has reached the edge of the cover slip. If the medium is not too rich abundant sporulation should have occurred by this time and spores and sporogenous cells will be adhering to the under side of the cover slip and around the agar on the surface of the slide. The cover slip is then removed and mounted in suitable mountant (dilute cotton blue in lactophenol) on a fresh slide. The agar is then removed from the culture slide and discarded and a drop of mountant placed in the centre. This is then covered with a cover slip and sealed.

The damp chamber is best made from a large 6 in. culture dish with filter paper in the bottom soaked in water or water plus 20 per cent glycerol. Two glass rods are placed on the filter paper to act as a rest for the slide or slides.

GENERAL GROWTH MEDIA

1. Potato Sucrose Agar
>   500 ml potato extract*
>   20 g sucrose
>   20 g agar
>   500 ml distilled water

The water and potato extract are mixed together and the sucrose and agar added. The mixture is heated slowly until the agar is dissolved and the pH adjusted if necessary to 6.5 with calcium carbonate. It is then dispensed in suitable bottles and autoclaved at 15 psi for 20 min.

*(Potato extract is prepared from 1800 g of mature main crop potatoes peeled and diced and suspended in muslin in 4500 ml of water and boiled for 10 min. The potatoes are then discarded and the liquor placed in large glass containers and autoclaved at 15 psi for 20 min. It can be stored in a refrigerator for use as required.)

2. Potato Dextrose Agar
>   200 g potato (scrubbed and diced)
>   15 g dextrose
>   20 g agar
>   1 litre water

New potatoes should be avoided. Boil potatoes for 1 hr and pass the mixture through a fine sieve, add agar and boil until dissolved, add dextrose and stir ; autoclave at 15 psi for 20 min.

3. Oatmeal Agar
>   30 g oatmeal (powdered)
>   20 g agar
>   1 litre water

Add oatmeal to water and gradually heat to boiling in a water bath or double saucepan and boil for 1 hr. Strain through muslin and make up liquor to 1 litre with water. Add agar and dissolve. Autoclave at 15 psi for 20 min.

SOIL ISOLATION MEDIA

4. Tapwater Agar
>   15 g agar
>   1 litre tapwater

Sterile wheat straw or rice grains may be added for use as a general medium. Many fusaria sporulate well on this.

5. Nash & Snyder's (1962) Peptone PCNB Medium
>   15 g Difco peptone
>   20 g agar
>   1 g potassium dihydrogen phosphate (KH_2PO_4)
>   0.5 g magnesium sulphate ($MgSO_4\ 7H_2O$)
>   1 g pentachloronitrobenzene (PCNB 75 per cent wettable powder)
>   300 ppm streptomycin (when cooled)*

*Lim (1974) added neomycin ($100\mu g/ml$) to this medium.

6. Papavizas' (1967) Peptone—PCNB modified
>   15 g Difco peptone
>   20 g agar
>   1 g potassium dihydrogen phosphate (KH_2PO_4)
>   0.5 magnesium sulphate ($MgSO_4\ 7H_2O$)
>   0.5 g PCNB—(Terraclor, a commercial product, is 75 per cent active)
>   0.5 g oxgall
>   100 mg streptomycin sulphate
>   50 mg chlortetracycline HCl

The last two components are thermolabile and should be added to the cooled agar.

MEDIA USED TO STIMULATE SPORULATION

7. CMC medium for stimulation of sporulation in *Fusarium graminearum* (Cappellini & Peterson, 1965)

   15 g carboxymethylcellulose (CMC 7MP—Hercules Powder Co.)
   1 g ammonium nitrate (NH_4NO_3)
   1 g potassium dihydrogen phosphate (KH_2PO_4)
   0.5 g magnesium sulphate ($MgSO_4$ $7H_2O$)
   1 g yeast extract
   1 litre distilled water

Used for shake cultures.

8. Bilay's medium modified by Joffe

   1 g potassium dihydrogen phosphate (KH_2PO_4)
   1 g potassium nitrate (KNO_3)
   0.5 g magnesium sulphate ($MgSO_4$ $7H_2O$)
   0.5 g potassium chloride (KCl)
   0.2 g starch powder
   0.2 g glucose
   0.2 g sucrose
   15 g agar
   1 litre water

Strips of pure cellulose lens paper were added before the agar had set.

MEDIA USED TO INCREASE INOCULUM

Many liquid media have been published : the following are frequently used.

9. Cerelose ammonium nitrate medium (Scheffer & Walker, 1953)

   50 g Cerelose
   10 g ammonium nitrate (NH_4NO_3)
   5 g potassium dihydrogen phosphate (KH_2PO_4)
   0.02 g ferric chloride ($FeCl_3$ $6H_2O$)
   1 litre water

10. Armstrong *Fusarium* medium

   20 g sucrose or glucose
   0.4 g magnesium sulphate ($MgSO_4$ $7H_2O$)
   1.6 g potassium chloride (KCl)
   1.1 g potassium dihydrogen phosphate (KH_2PO_4)
   5.9 g calcium nitrate ($Ca(NO_3)_2$)
   Ferric chloride ($FeCl_3$)
   Manganese sulphate ($MnSO_4$) } 0.22 ppm of each
   Zinc Sulphate ($ZnSO_4$)

11. Medium used to study staling of *Fusarium oxysporum* (Park, 1961)

   0.7 g glucose
   0.5 g magnesium sulphate ($MgSO_4$ $7H_2O$)
   0.2 g potassium dihydrogen phosphate (KH_2PO_4)
   0.1 g ammonium nitrate (NH_4NO_3)
   15 g agar
   1 litre distilled water

MEDIA FOR PERITHECIAL PRODUCTION

12. Wheat or rice straw or plant stems in petri-dishes or longer pieces standing in tap water agar or in water in large medical flats or other suitable containers.

TABLE I

Surface sterilizing agent	Concentration (per cent)	Time (min)	Rinsing agent
Formaldehyde in sol.	51	1-5	70 per cent ethyl alcohol then sterile water
Hydrogen peroxide	3	1-5	Sterile water
Potassium permanganate	2	1-5	Sterile water
Sodium or calcium hypochlorite*	0.35	1-5	Sterile water
Mercuric chloride**	0.001	1-5	70 per cent ethyl alcohol then sterile water
Ethyl alcohol	75	1-5	Sterile water
Silver nitrate	1	1-5	Sterile sodium chloride then sterile water

*Normally a 1-10 dilution of commercial solution (Chlorox) is sufficient.

**Usually made up from stock solution ($HgCl_2$ 20g and conc. HCl to make 100 ml) 5 ml and 995 ml water.

GROWTH RATE

Determinations of growth rates follow Gordon (1952) and represent the average colony diameter of a number of single conidial isolates on potato sucrose agar (pH 6.5) maintained at 25°C for four days.

PIGMENTATION

All colour terms used in this text are defined in relationship to those used by Rayner (1970). The cultures were exposed 12-14 inches below two day light fluorescent tubes. This exposure is not critical and exposure to normal summer daylight is adequate for purposes of determination.

PHOTOGRAPHS AND LINE DRAWINGS

All photographs were taken from water mounts or mounts of lactophenol and dilute cotton blue ; all are printed × 800 unless otherwise stated. Line drawings were made by use of a Leitz drawing apparatus and are × 1800.

ACKNOWLEDGMENT

I wish to thank Mrs. G.B. Butterfill for continued technical help, Mr. D. W. Fry for producing the photographs and Mrs. M. S. Rainbow for kindly typing the manuscript. In particular I wish to thank the many contributors who have over the years supplied thousands of cultures and specimens.

(1) FUSARIUM OXYSPORUM Schlecht., *Flora Berol.* 2 : 139, 1824

Growth rate 4.5 cm.

Culture pigmentation white, peach, salmon, vinaceous grey to purple, violet.

Microconidia oval-ellipsoid, cylindrical, straight or curved, 5-12 × 2.2-3.5μ, produced
from simple short lateral phialides.

Macroconidia generally 3-5 septate, 27-60 × 3-5μ.

Chlamydospores globose, formed singly or in pairs, intercalary or on short lateral
branches.
Chromosome number = 12.

Perithecial state ? *Gibberella* not confirmed.

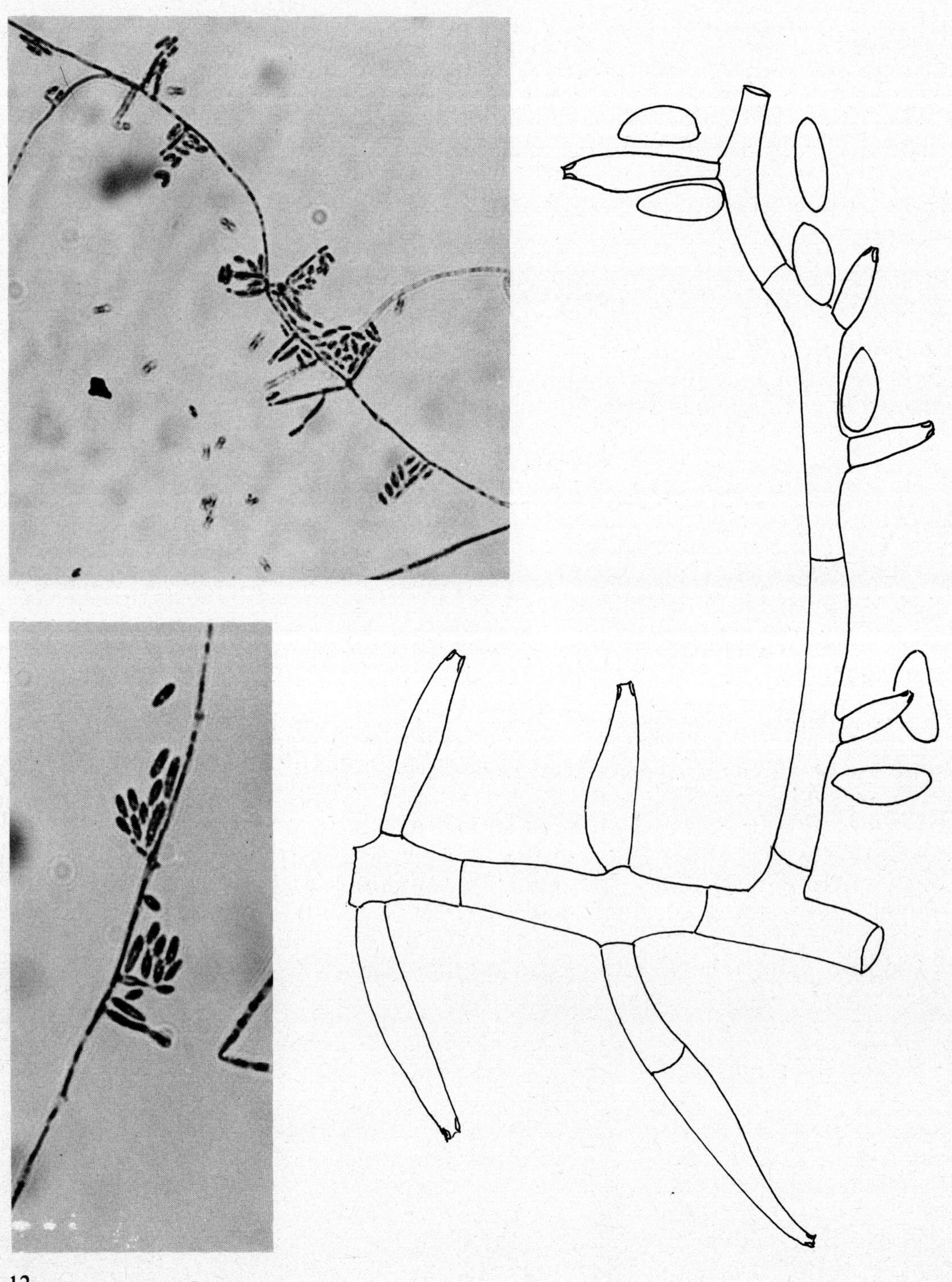

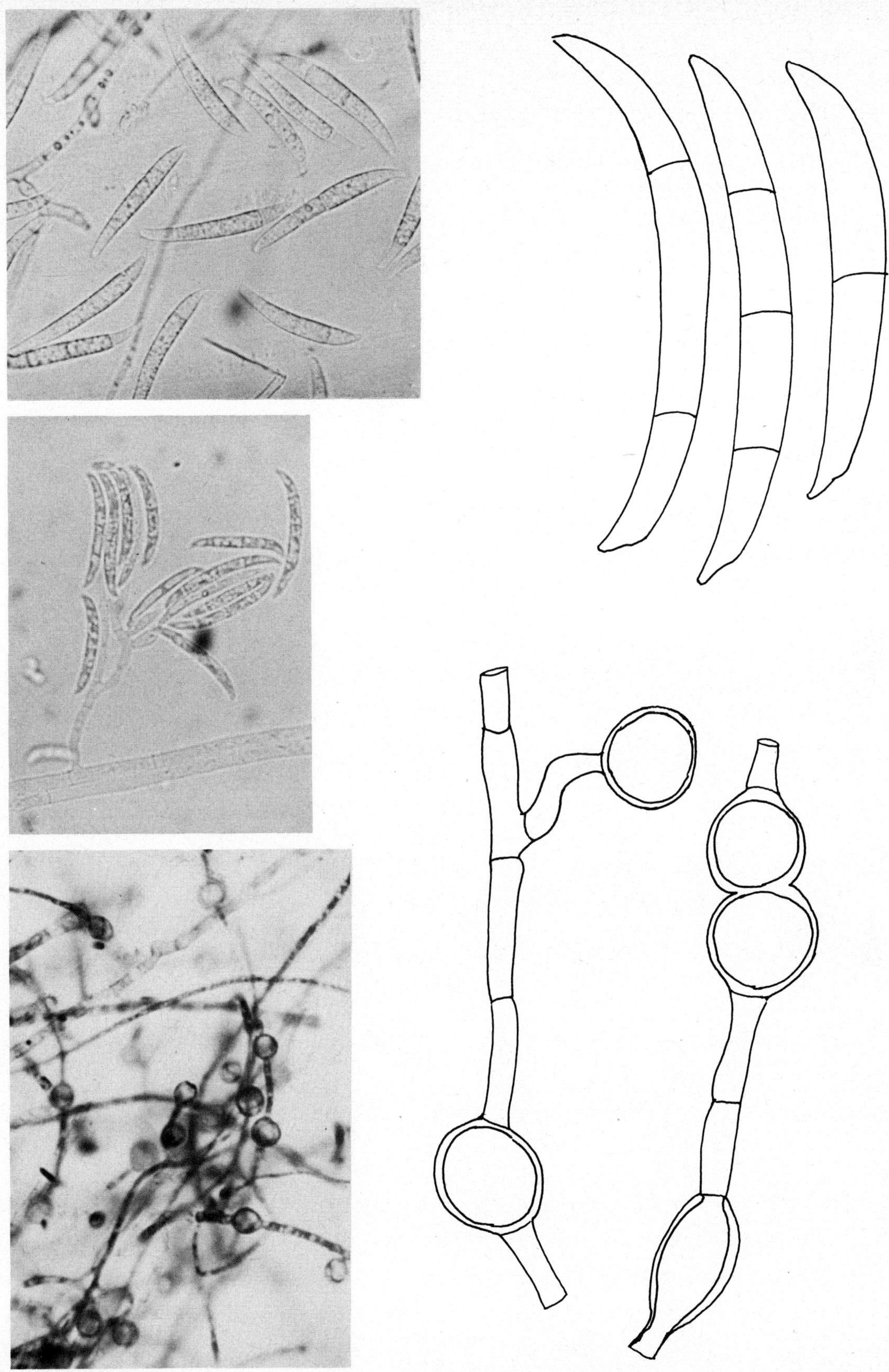

Diagnostic characters : The short simple phialides producing the microconidia separate it from *F. solani* and together with presence of chlamydospores from *F. moniliforme* and its varieties.

(2) FUSARIUM SOLANI (Mart.) Sacc., *Michelia* **2** : 296, 1881

Growth rate 3.2 cm.

Culture pigmentation greyish-white to blue or bluish-brown.

Microconidia 8-16 × 2-4μ, cylindrical to oval and may become 1- septate, produced from long lateral phialides 45-80 × 2.5-3μ, laterally borne or on branched conidiophores.

Macroconidia from the side inequilaterally fusoid with widest point above the centre, length variable in different strains, i.e. 1-5 septate, 35-55 × 4.5-6μ, 5-9 septate, 35-55 × 4.5-6μ, 5-9 septate, 45-100 × 5-8μ.

Chlamydospores globose, smooth to rough walled, 9-12 × 8-10μ borne singly or in pairs on short lateral branches or intercalary.

Chromosome number = 8.

Perithecial state *Nectria haematococca* Berk. & Br.

Ascospores ellipsoid to ovate, 11-18 × 4-7μ, becoming light brown with longitudinal striations.

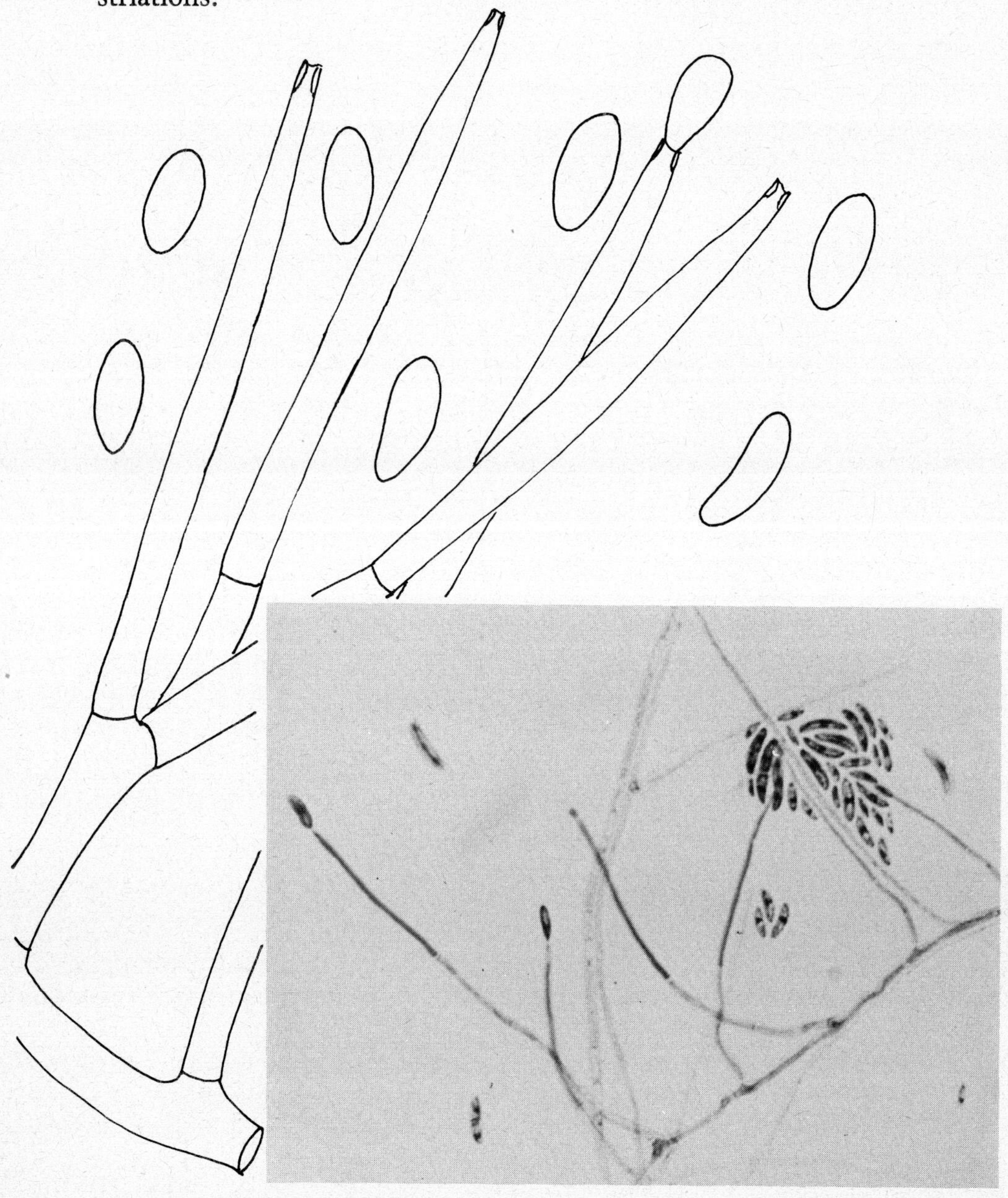

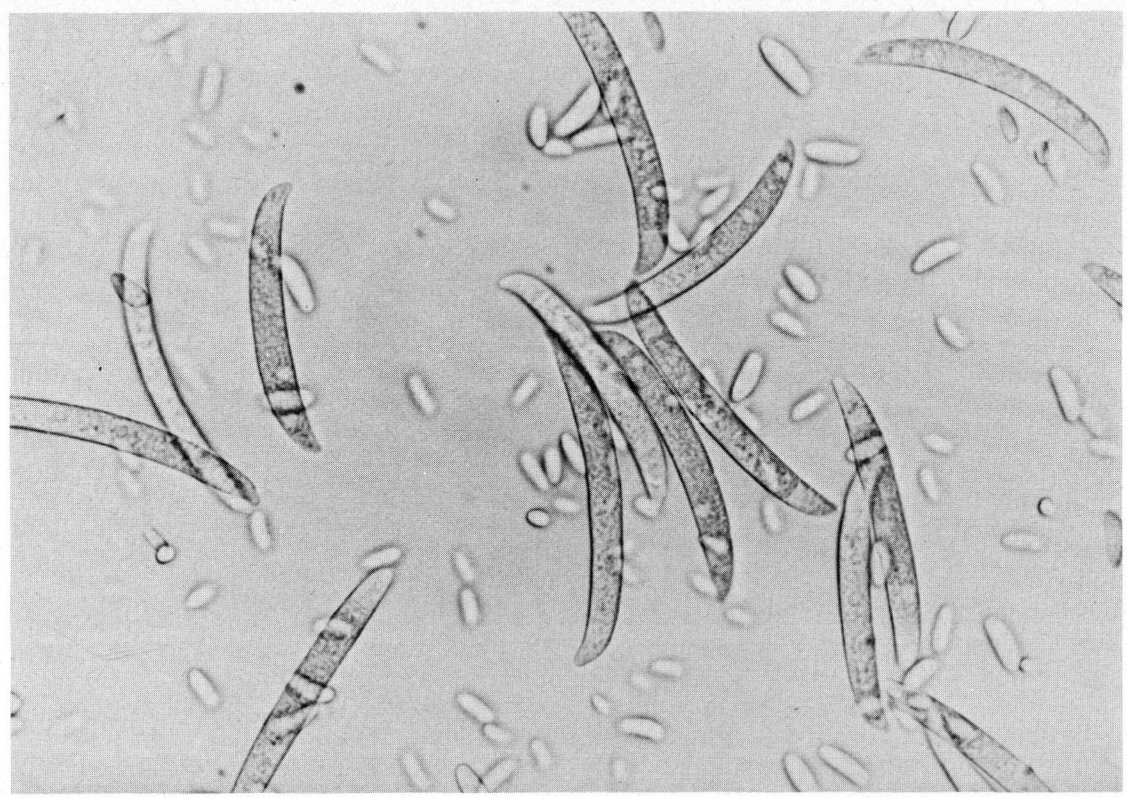

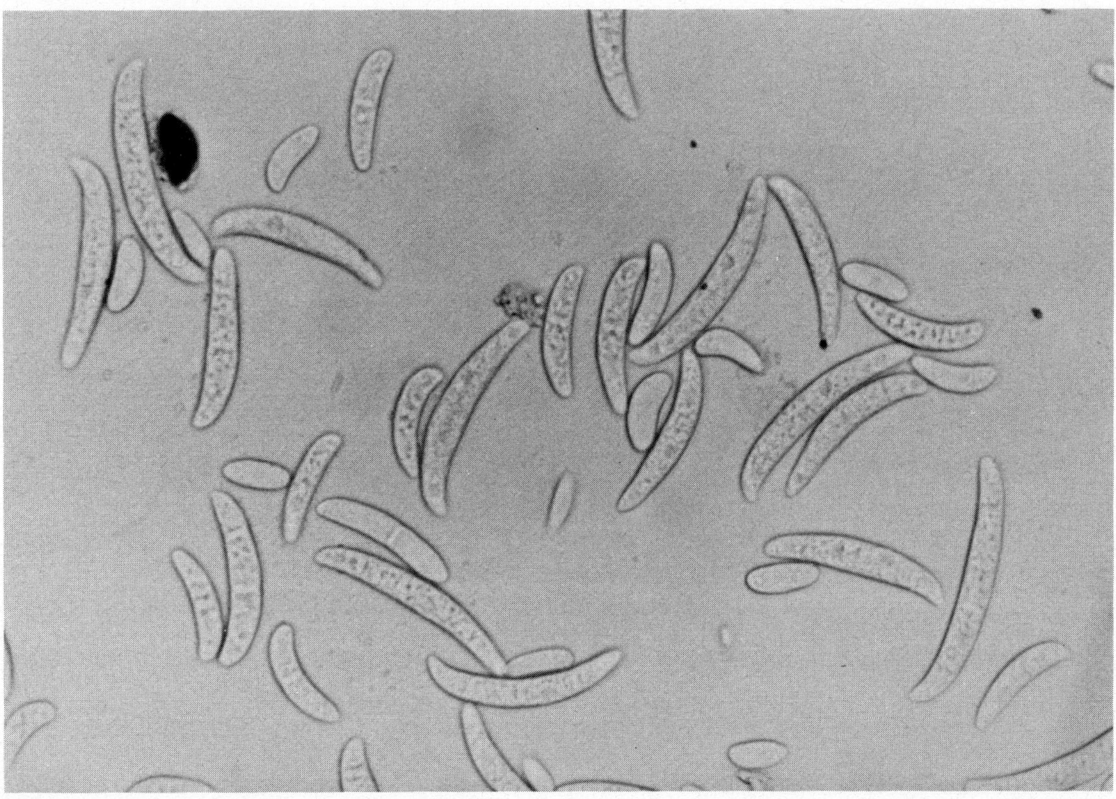

Diagnostic characters distinguishing it from *F. oxysporum* are the long phialides and the branched often elaborate microconidiophores and the shape of the macroconidia. *F. solani* var. *coeruleum* causes powdery rot of potatoes in storage. It usually has a strong dark violet-blue pigmentation, reduced microconidial formation and the occasional production of polyphialides producing the macroconidia.

(3) FUSARIUM MONILIFORME Sheldon, *Rep. Neb. agric. Exp. Stn* **17** : 23-32
1904

Growth rate 4.6 cm.

Culture pigmentation : peach salmon, vinaceous purple to violet.

Microconidia fusoid to clavate, 5-12 × 1.5-2.5μ, occasionally becoming 1- septate
and produced in chains from subulate lateral phialides, 20-30μ long by 2.3μ
at the base.

Macroconidia : some strains do not readily form macroconidia but when present
they are in equilaterally fusoid, thin walled, 3-7 septate, 25-60 × 2.5-4μ.

Chlamydospores absent but globose stromatic initial cells may be present in some
cultures.

Chromosome number = 7.

Perithecial state *Gibberella fujikuroi* (Sawada) Wollenw. Ascospores 1-3 septate,
14-18 × 4.5-6μ.

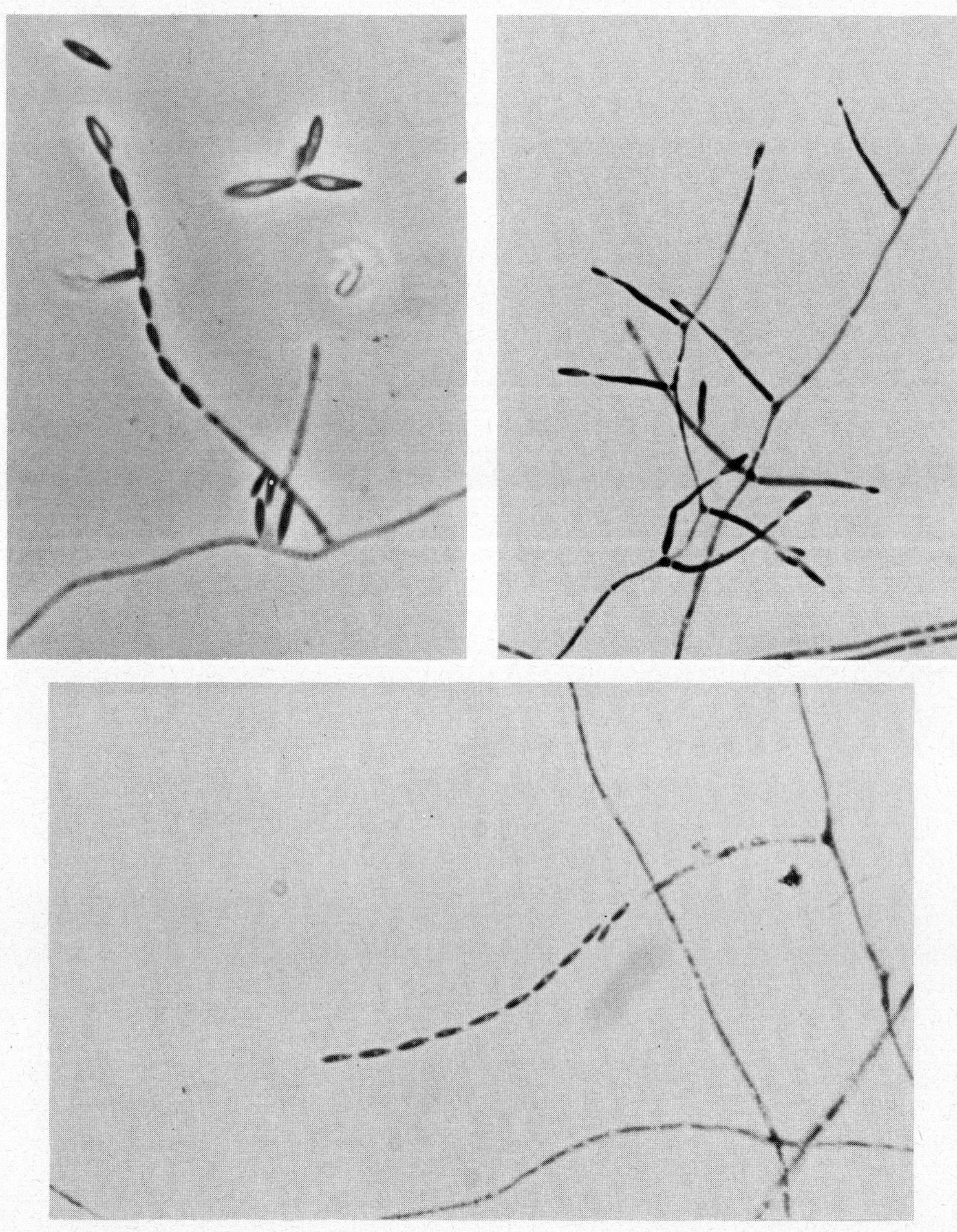

16

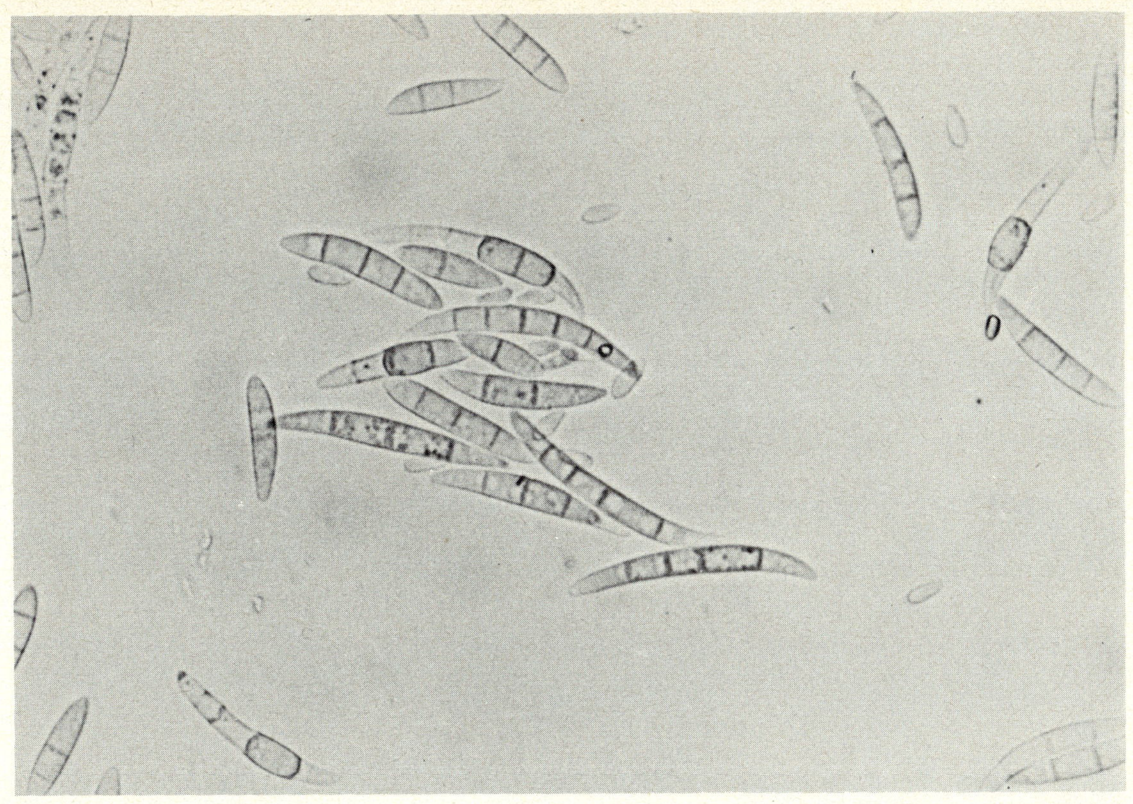

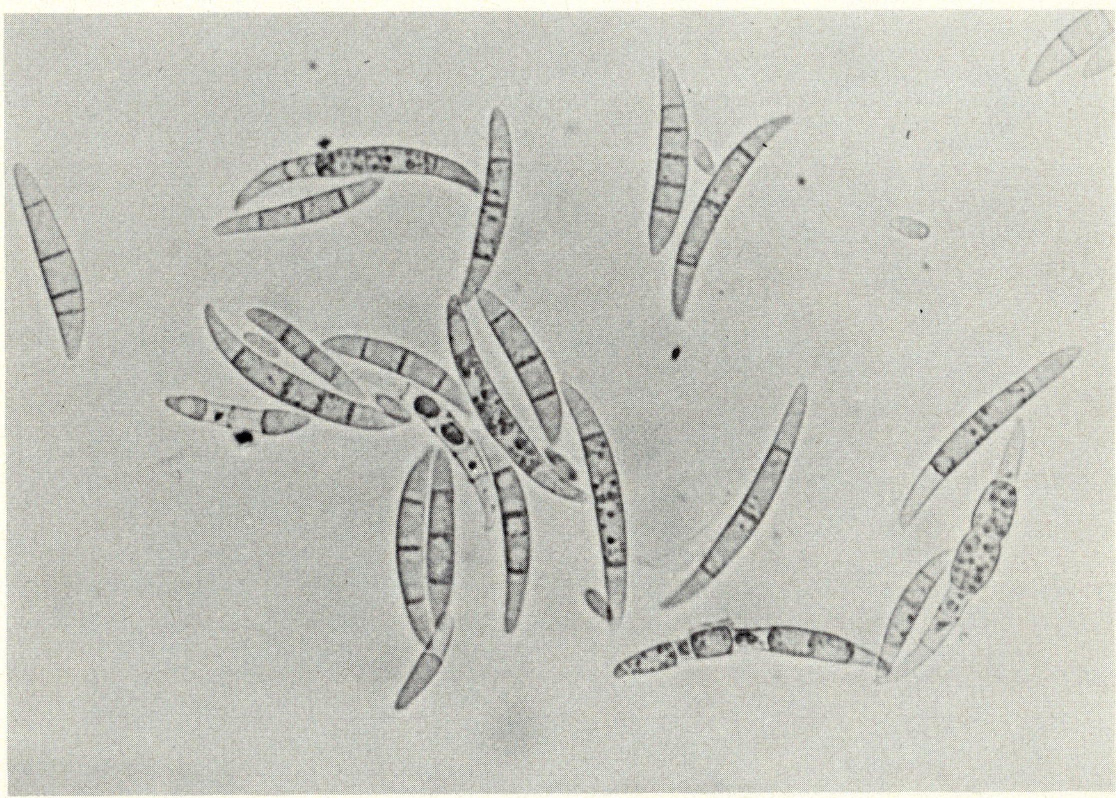

Diagnostic characters : This species has similar pigmentation to *F. oxysporum* but it is readily separated from this species by the chains of microconidia which can be best observed *in situ* in a plate culture under the low power of a compound microscope. The absences of chlamydospores is also a point of separation.

(4) FUSARIUM MONILIFORME var. SUBGLUTINANS Wollenw. & Reink.
Phytopathology **15** : 163, 1925

Growth rate 4.5 cm.

Culture appearance and pigmentation is the same as *F. moniliforme*.

Microconidia form from polyphialides and accumulate in globose heads, chains not produced ; they are oval to obclavate, 8-12 × 2.5-3μ.

Macroconidia when present formed from simple phialides; they are 3-5 septate and measure 32-53 × 3-4.5μ.

Chlamydospores not produced.

Perithecial state *Gibberella fujikuroi* var. *subglutinans* Edwards

Ascospores 1-3 septate, 12-15 × 4.5-5μ.

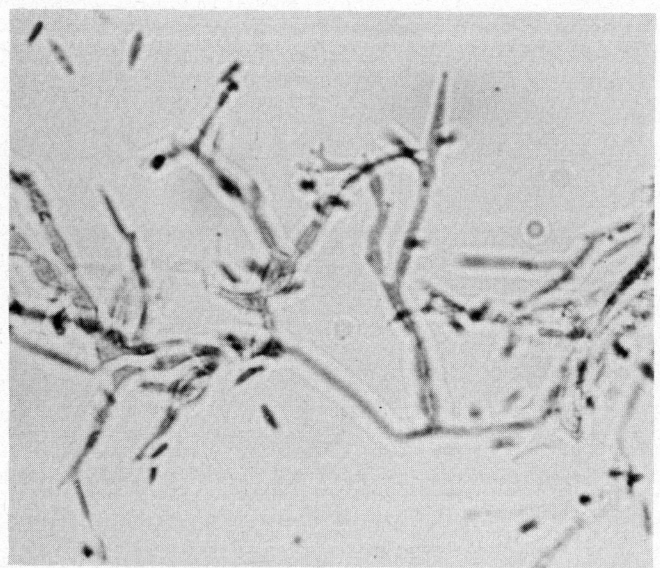

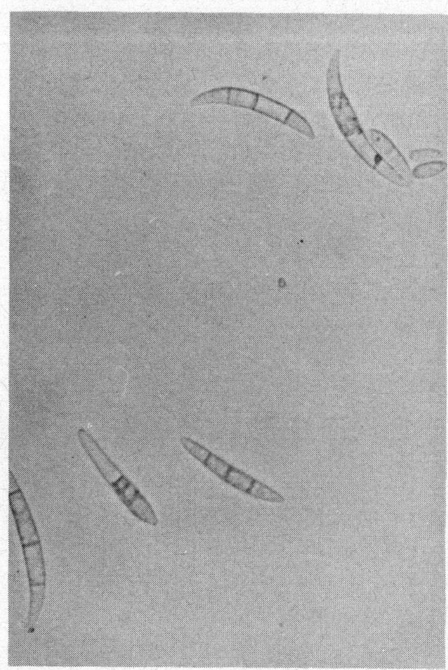

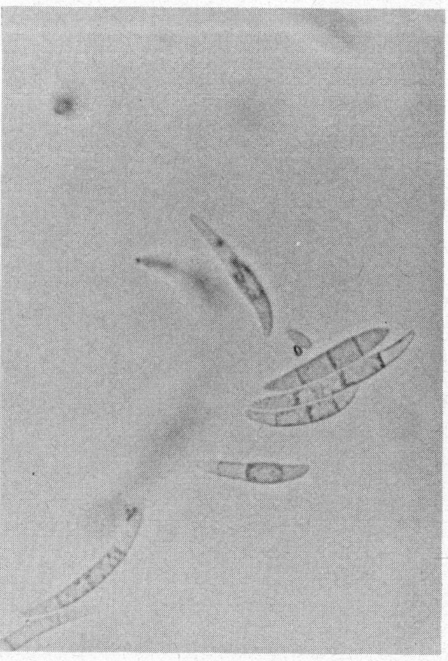

Diagnostic characters : The presence of polyphialides producing uniform microconidia is a readily observable diagnostic character.

(5) FUSARIUM MONILIFORME var. ANTHOPHILUM(A. Braun) Wollenw.
   Z. Parasitenk. **3** : 397-399, 1931.

Growth rate 4.5 cm.

Cultural characteristics are the same as for *F. moniliforme.*

Microconidia formed both from simple phialides and polyphialides, two distinct types
   are present ; a, fusoid to allantoid, 6-9 $\times$ 2-3μ ; b, oval to globose, 5-5.8 $\times$
   3.5-5.5μ.

Macroconidia when present are curved, fusoid, 3-5 septate, 30-50 $\times$ 3-4μ.

Chlamydospores absent.

Perithecial state unknown.

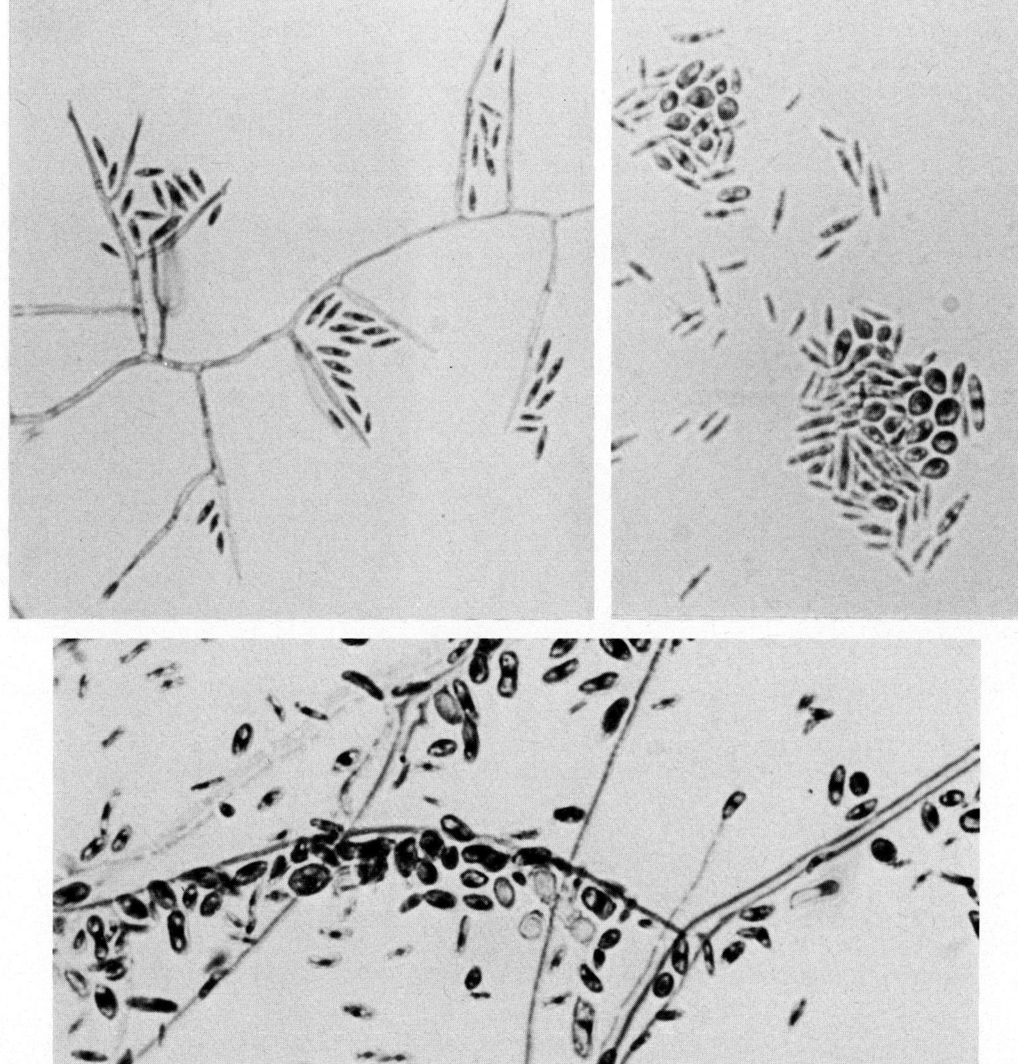

Diagnostic characters : The presence of two morphologically distinct microconidial
forms readily separates this variety from *F. moniliforme* var. *subglutinans* and the presence of polyphialides separates both varieties from the parent strain.

(6) FUSARIUM DECEMCELLULARE Brick, *Jber. ver. Angew Bot.* **6** : 227, 1908

Growth rate 3.2 cm.

Cultural pigmentation : rose darkening to red, aerial mycelium white but pustules of macroconidia cream to yellow.

Microconidia formed in chains from well developed phialides. They are oval, aseptate to 1-septate, 10-15 × 3-5μ.

Macroconidia formed on sporodochia from well developed phialides ; they are 7-10 septate, 55-130 × 6-10μ.

Chiamydospores absent.

Chromosome number = 7.

Perithecial state *Calonectria rigidiuscula* Berk & Br. Ascospores 3 septate, 22-28 × 7-10μ with longitudinal striations.

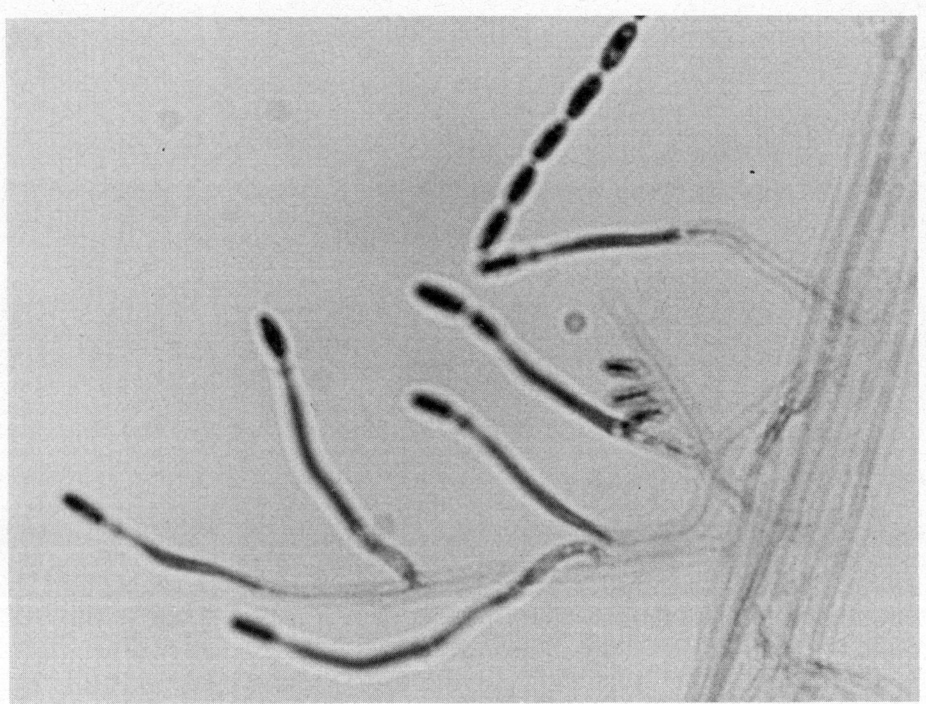

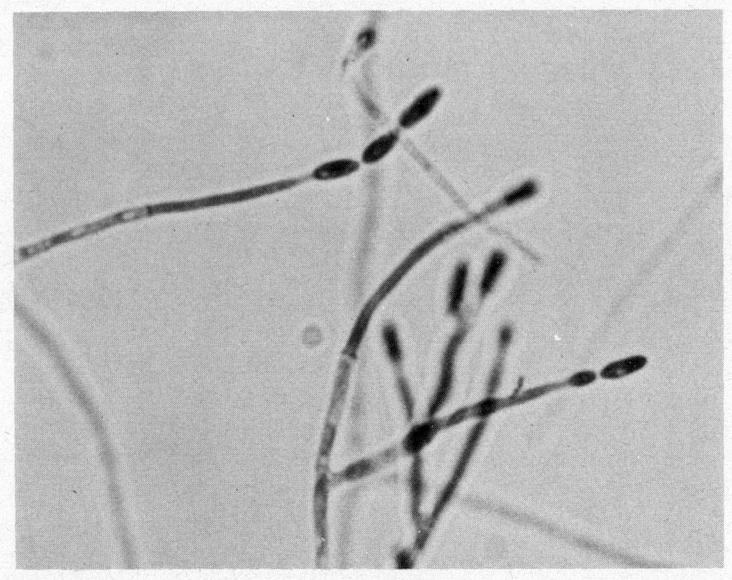

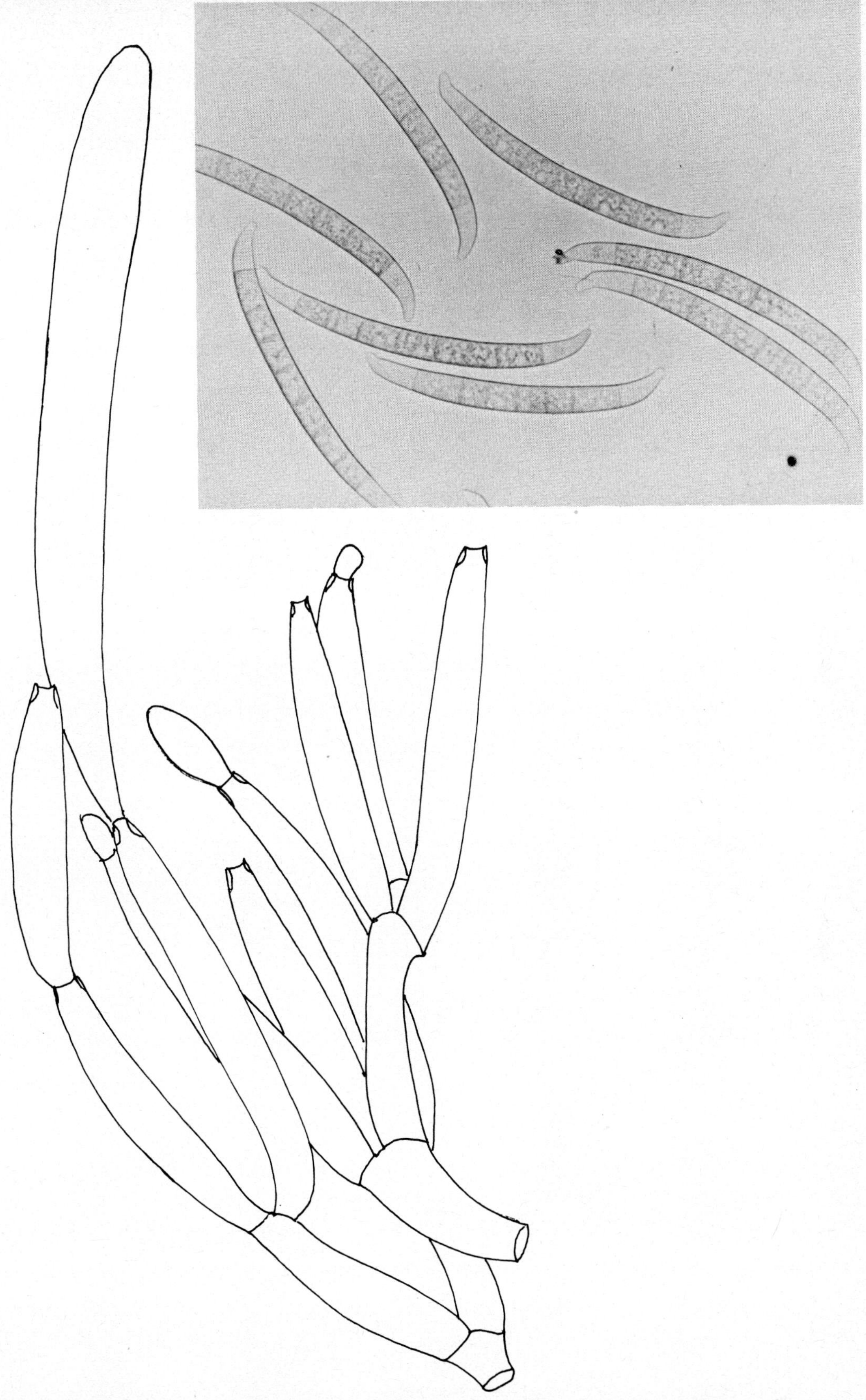

Diagnostic character : *F. moniliforme* and *F. decemcellulare* are the only two *Fusarium* species producing microconidia in chains. Apart from this common factor they are distinct in almost every other character, spore shape and size, pigmentation, perithecial form.

(7) FUSARIUM POAE (Peck) Wollenw. in Lewis, *Bull. Me agric. Exp. Stn* **219** : 254-258, 1913

Growth rate 7.6 cm.

Culture pigmentation : white, yellow, salmon to livid-red or vinaceous, colours somewhat unstable and some strains remain colourless.

Microconidia ampulliform, 8-12 $\times$ 7-10μ, to globose, 7-10μ diam., formed from lateral pores or small globose to obclavate phialides.

Macroconidia when present 3-septate, 20-40 $\times$ 3-4.5μ.

Chlamydospores globose, formed sparsely, singly, or in chains in older cultures.

Chromosome number = 6.

Perithecial state unknown.

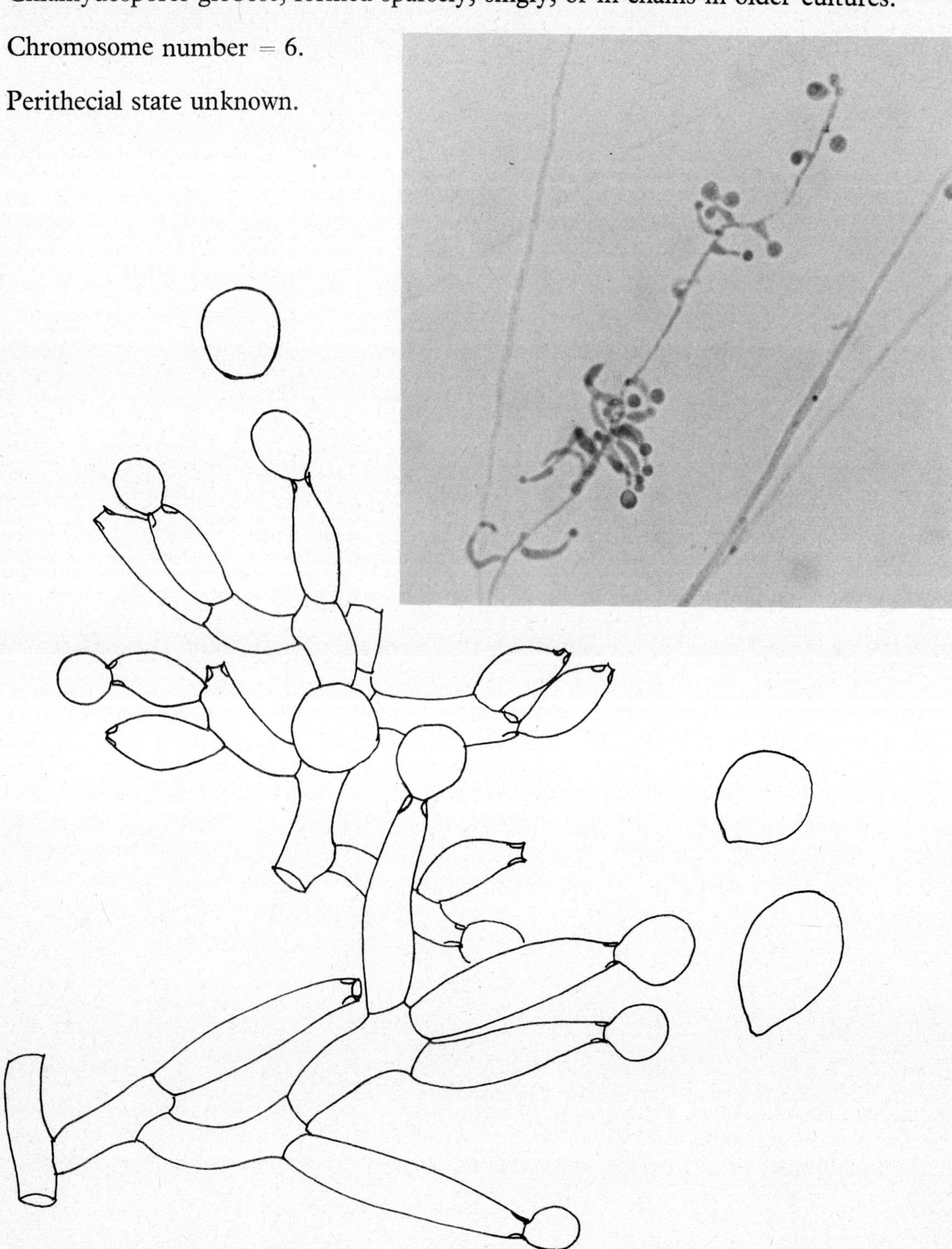

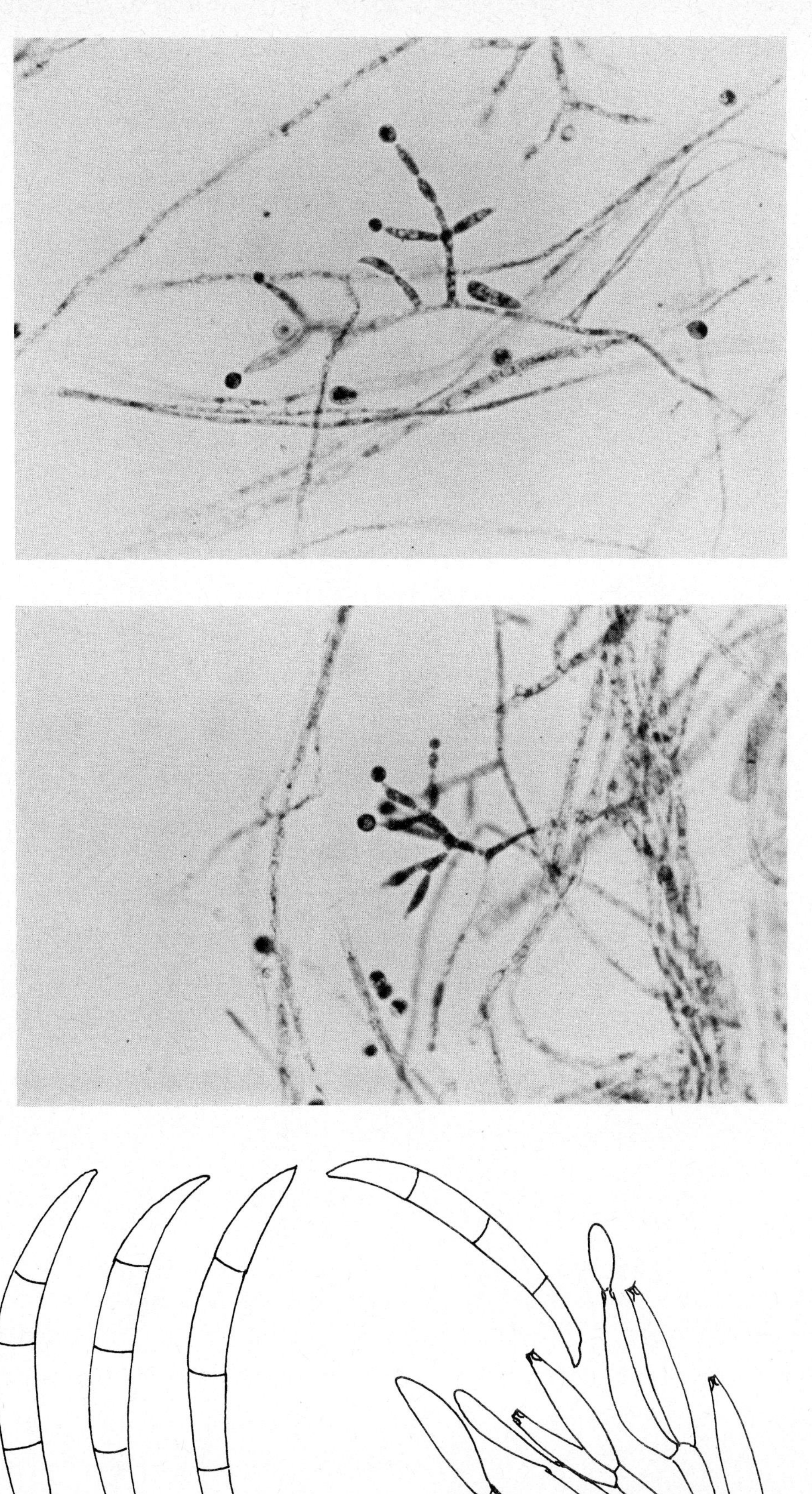

Diagnostic characters : The globose microconidia formed from phialides are specific to this species. This high growth rate also separates it from the related *F. tricinctum*.

(8) FUSARIUM TRICINCTUM (Corda) Sacc., *Sylloge Fung.* **4** : 700, 1886

Growth rate 4.0 cm.

Culture pigmentation carmine, red to purple.

Microconidia ovate to pyriform, 7-14 × 4.5-7.5μ, becoming 1-septate and produced
by subulate phialides.

Macroconidia 3-5 septate, 24-50 × 3.5-4.6μ.

Chlamydospores when present, globose, 10-12μ diam., intercalary, single or in chains,
or occasionally borne on short lateral branches.

Perithecial state unknown.

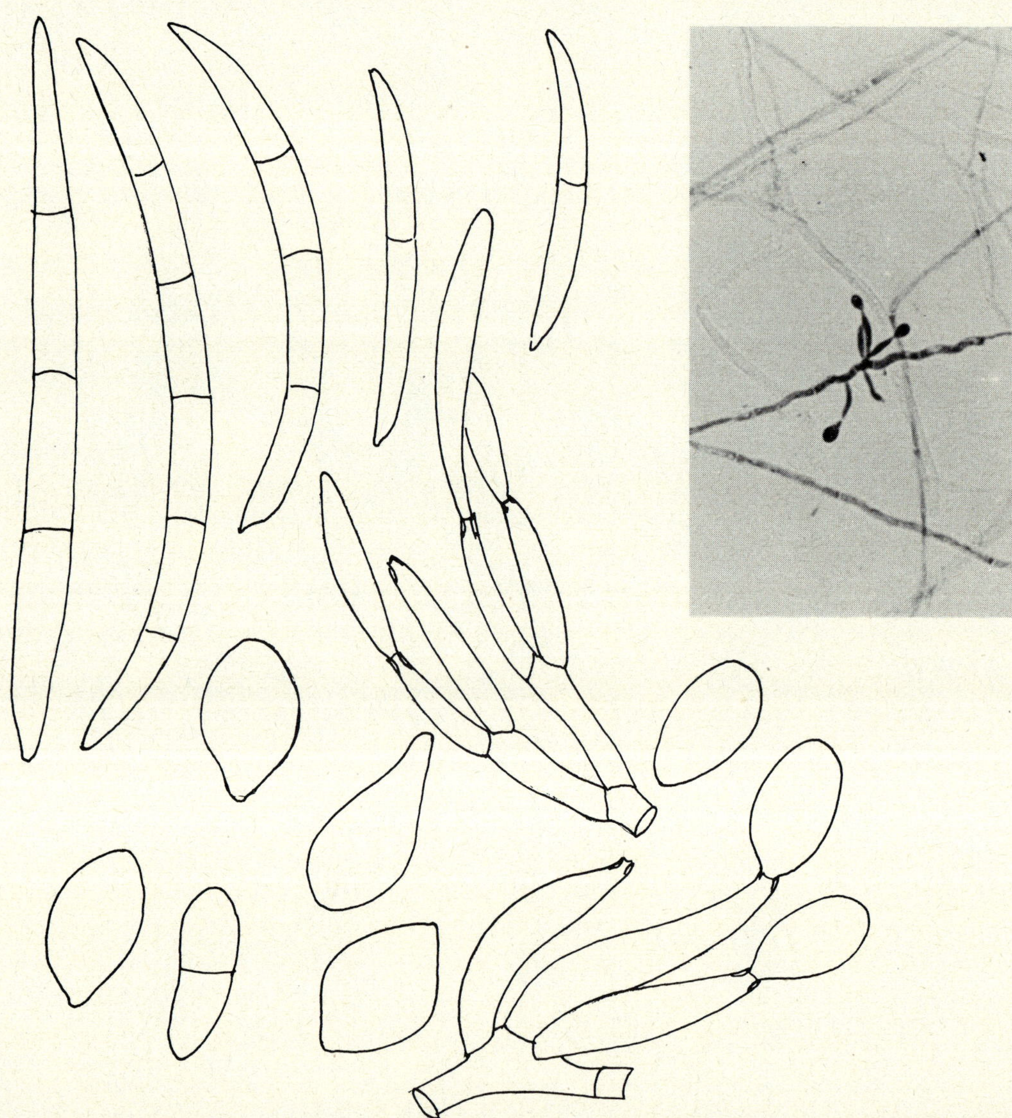

Diagnostic characters are the red pigmentation and the pyriform to obovoid microconidia produced from simple phialides. Although similar to *F. poae*, which has more
globose microconidia, it also has a much slower growth rate.

(9) FUSARIUM EQUISETI (Corda) Sacc., *Sylloge Fung.* **4** : 707-708, 1886

Growth rate 5.9 cm.

Culture pigmentation : peach usually changing to avellaneous and finally becoming buff brown.

Macroconidia only are produced and these may be variable in size and are produced from single solitary or grouped phialides; conidia 4-7 septate, 22-60 × 3.5-9μ, often with elongated apical cell.

Chlamydospores globose, 7-9μ diam., intercalary, solitary in chains or clumps.

Chromosome number = 8.

Perithecial state *Gibberella intricans* Wollenw. Ascospores 1-3 septate, 21-33 × 4-5.5μ.

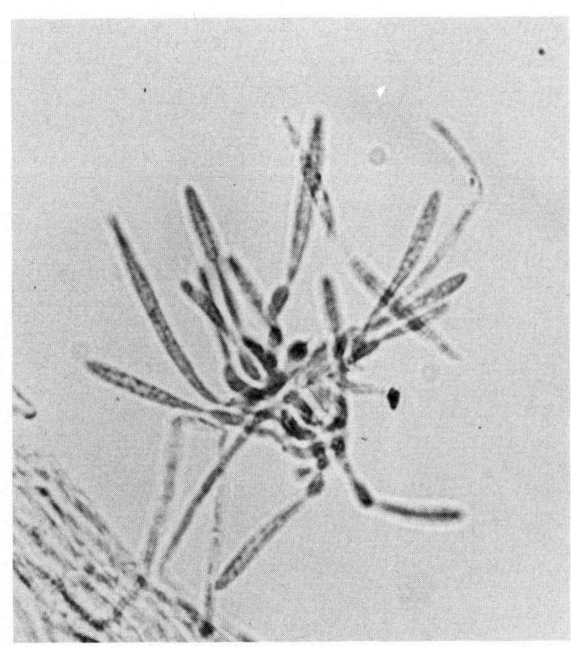

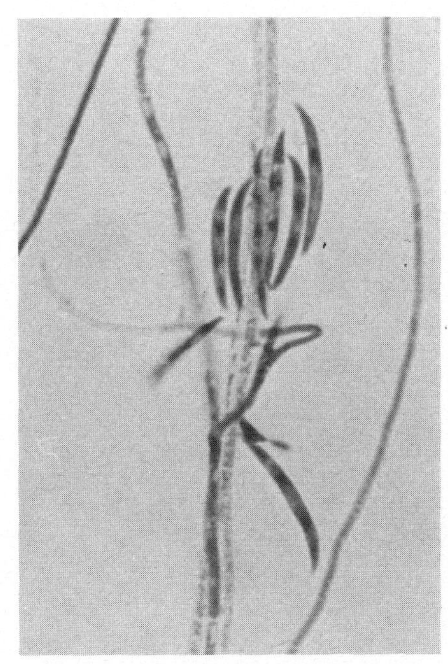

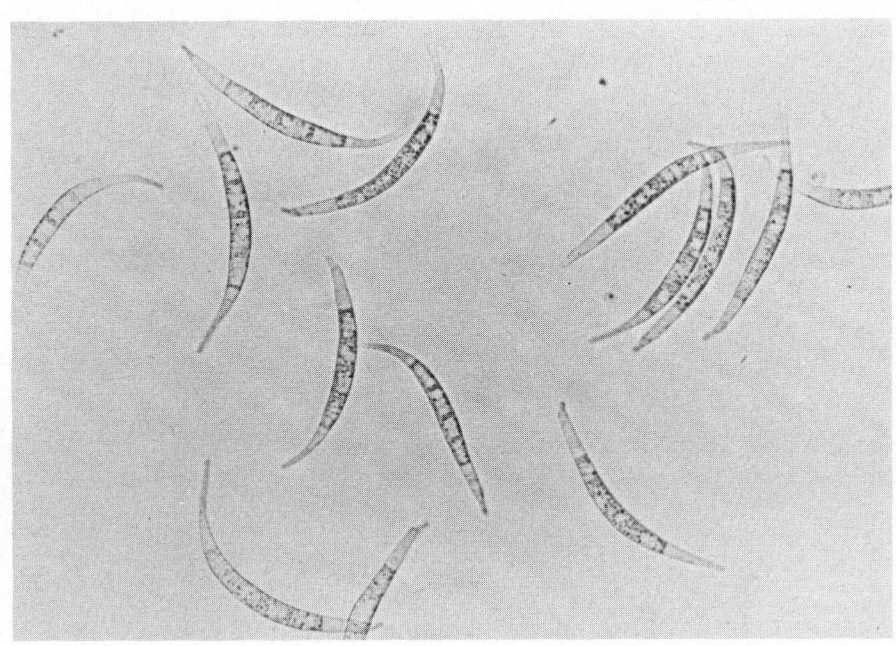

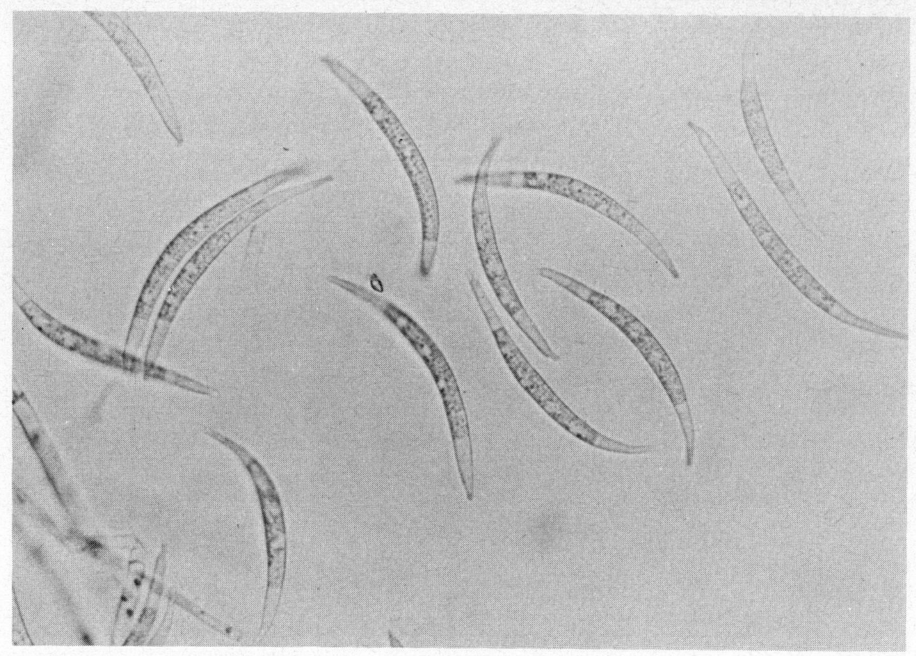

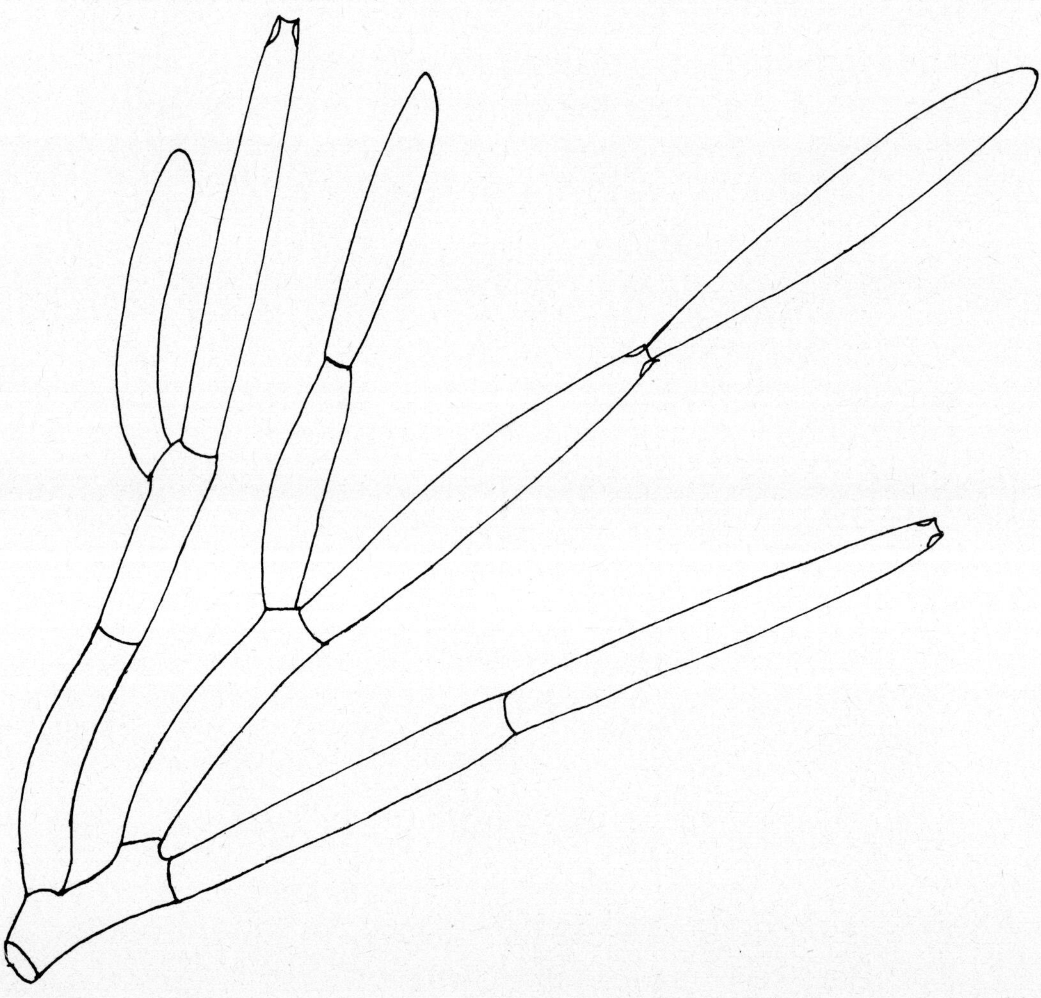

Diagnostic characters : The sequence of pigmentation is similar to *F. semitectum* but no blastospores are produced. Pigmentation readily separates it from *F. acuminatum.*

(10) FUSARIUM ACUMINATUM Ellis & Everhart, *Proc. Acad. Sci. Philadelphia*
441, 1895.

Growth rate 4.5 cm.

Culture pigmentation : saffron to bay to carmine red.

Macroconidia only are produced ; they are variable in different isolates, 3-7 septate,
30-70 × 3.5-5μ often with an incurved elongation of the apical cell, and
are produced from phialides.

Chlamydospores are intercalary in knots or chains.

Perithecial state *Gibberella acuminata* Wr. Heterothallic, ascospores 13-22 × 6-7μ.

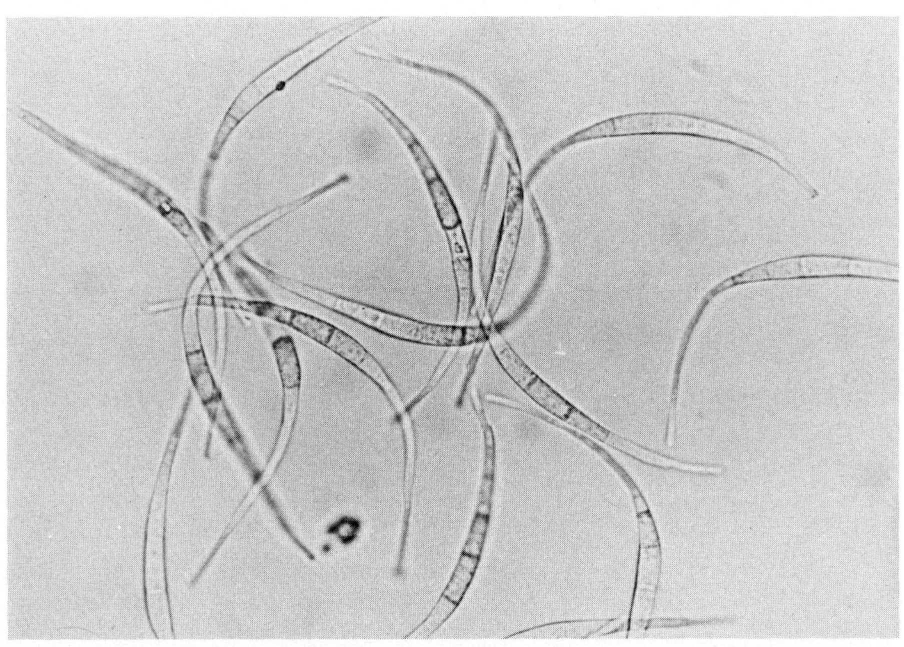

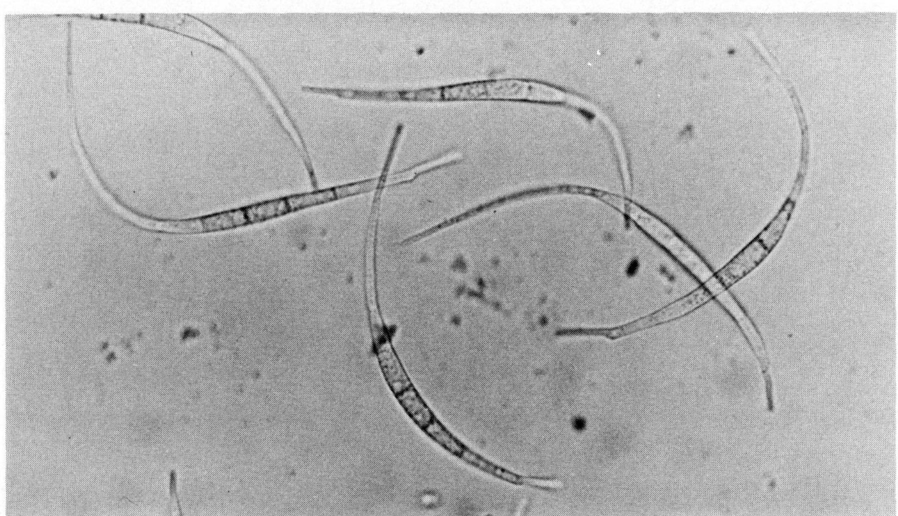

Diagnostic characters : Varation in spore form is similar to *F. equiseti* but carmine
red pigmentation is quite distinct.

(11) FUSARIUM LATERITIUM Nees, Syst. Pilze Schwamme p. 31, 1817 : Link.
Spec. Plant. 6 (2) : 106, 1824

Growth rate = 2.8 cm.

Cultural pigmentation variable peach to deep orange, vinaceous to reddish brown,
greenish yellow to blue-black.

Macroconidia only produced generally in sporodochia are 3-7 septate, beaked at the
apex, 22-50 × 3.5-5.5μ, formed from simple phialides.

Chlamydospores form sparsely and are oval to globose, 7-10 × 7μ or 7-8μ diam.

Perithecial state : *Gibberella baccata* (Wallr.) Sacc.; ascospores 12-18 × 4.5-7.5μ.

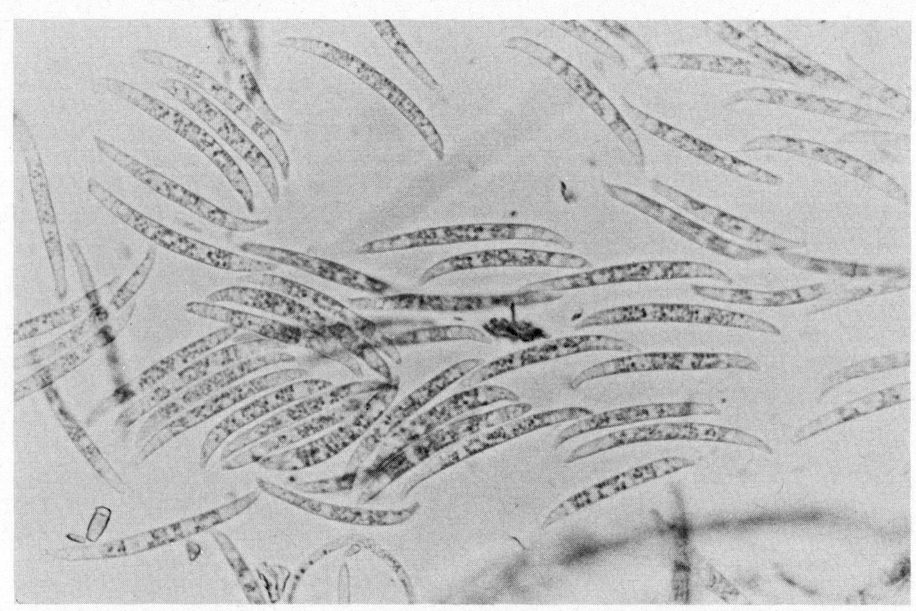

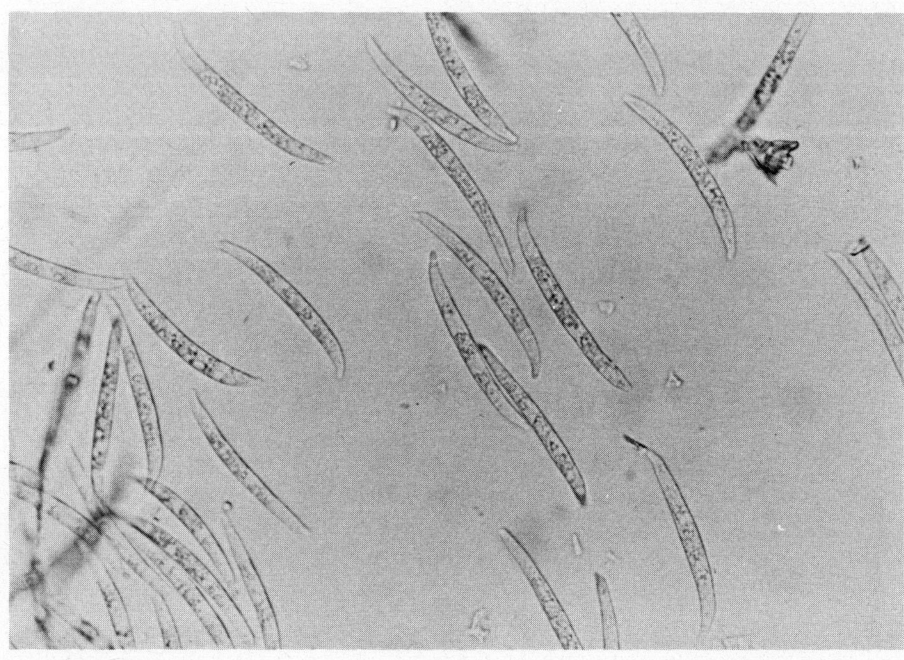

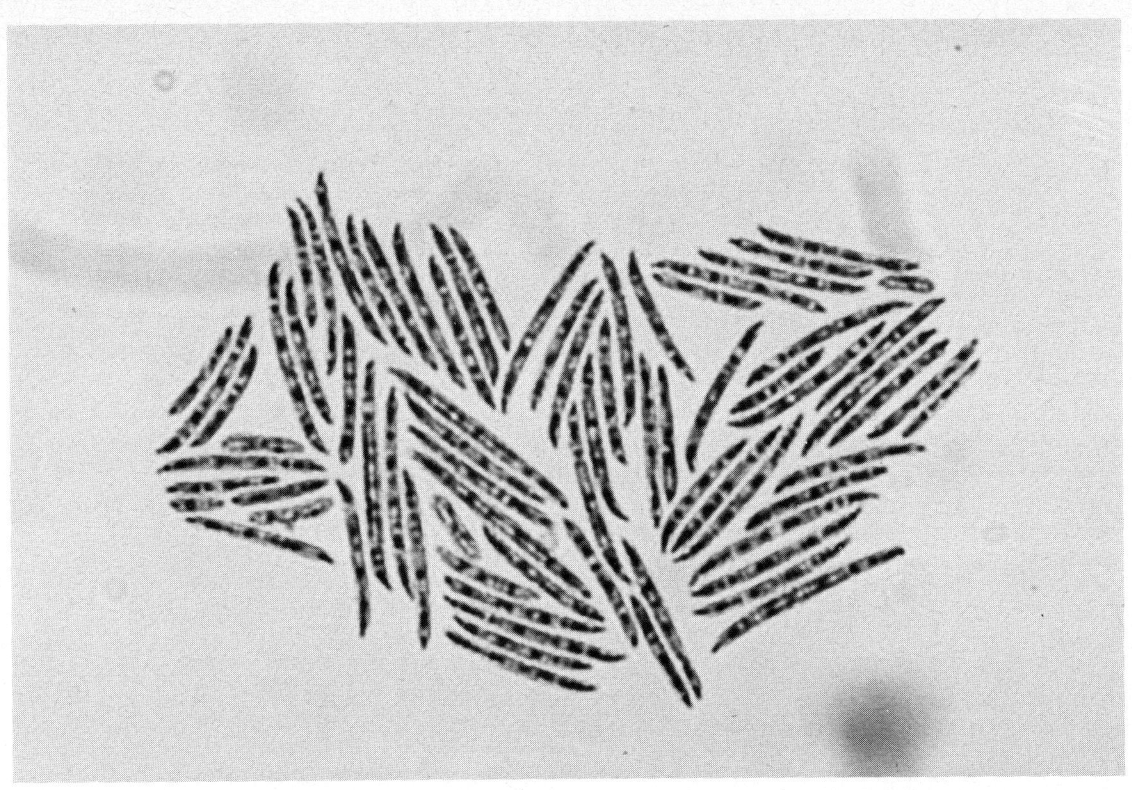

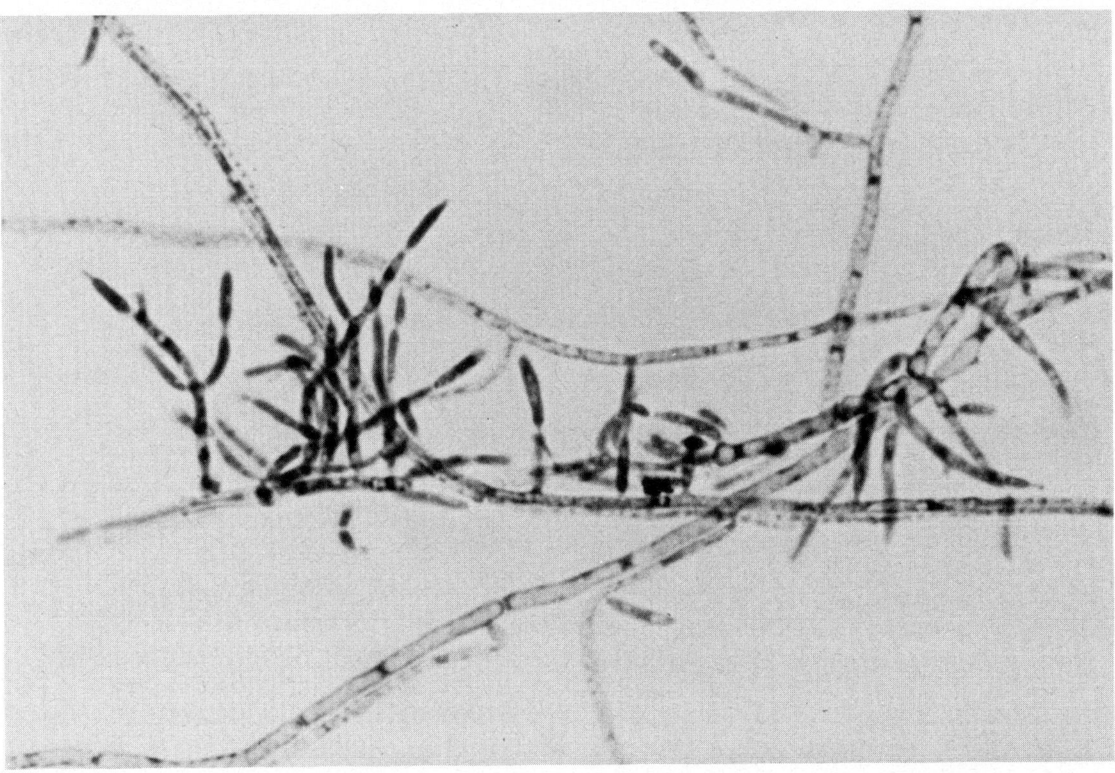

Diagnostic characters : Straight beaked macroconidia, pigmentation, perithecial state.

(12) FUSARIUM STILBOIDES Wollenw., *Fus. auto. del.* 615, 1924.

Growth rate 3.0 cm.

Cultural pigmentation : characteristic carmine red with white floccose mycelium becomes reddish-brown later.

Macroconidia only produced; they are 3-7 septate, 20-82 × 3-4·5µ, formed from simple phialides or borne in sporodochia.

Chlamydospores formed sparsely both intercalary and on short lateral branches; they are globose, 8-13µ diam.

Perithecial state *Gibberella stilboides* Gordon ex Booth ; ascospores 1-3 septate, 12-18 × 4-5.5µ.

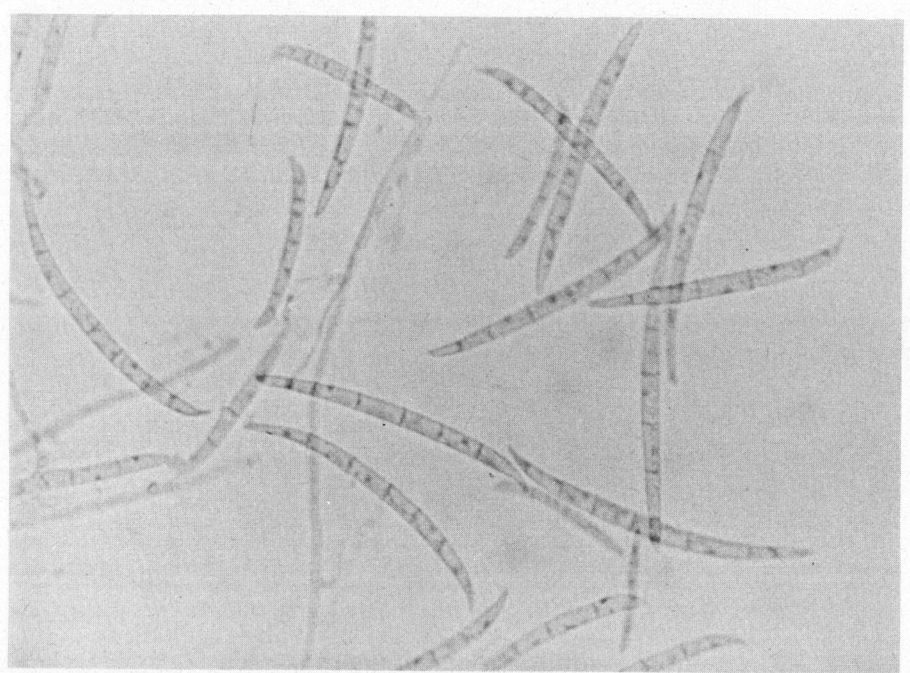

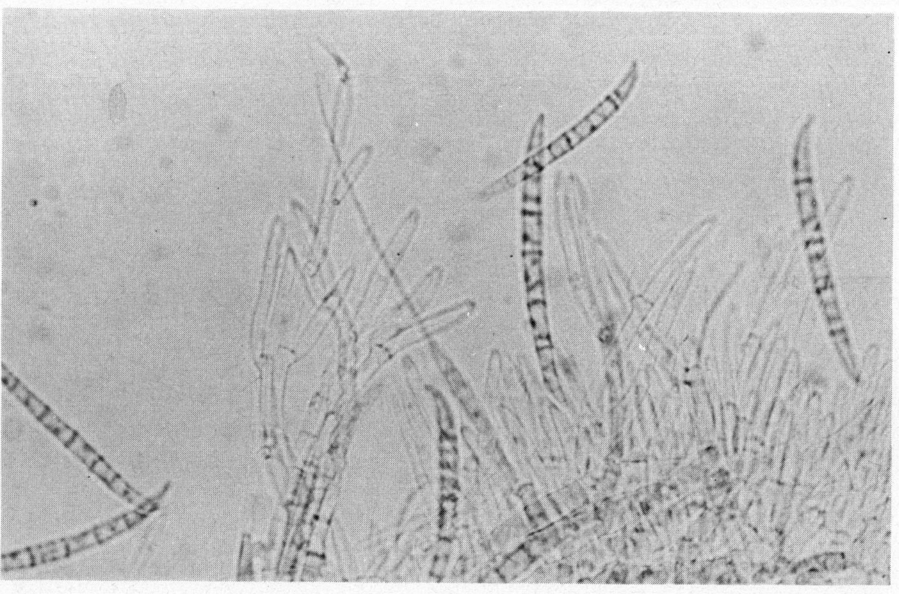

Diagnostic characters : The appearance of the species in culture, its spore form and its association with scaly bark disease and dieback of coffee are characteristic features.

30

(13) FUSARIUM UDUM Butler, *Mem. Dep. Agric. India, Bot.* ser. 2 (9) : 54, 1910.

Growth rate 4.2 cm.

Culture pigmentation : pale sulphureus to rose buff becoming salmon orange with production of conidia, occasional strains have purple pigmentation.

Conidia variable with a strongly curved or hooked apex, 6-8 × 3-3.5μ and 30-40 × 3-3.5μ but with no clear distinction between micro and macroconidia.

Chlamydospores often sparse, oval to globose, 8-11 × 8-12μ.

Perithecial state unknown.

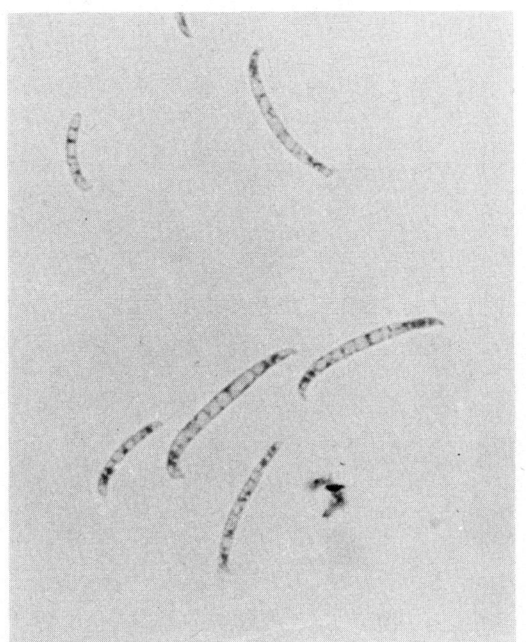 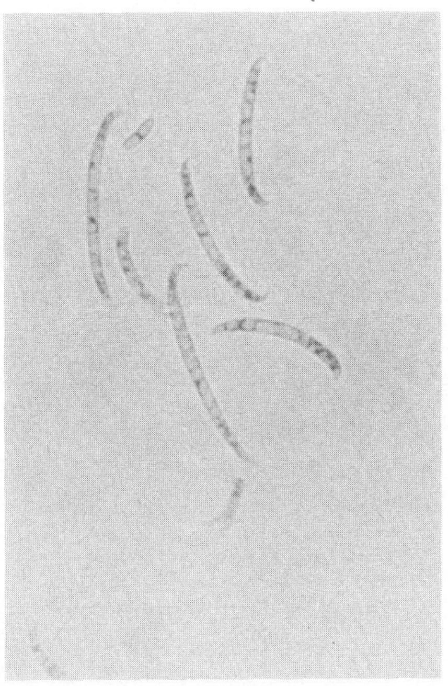

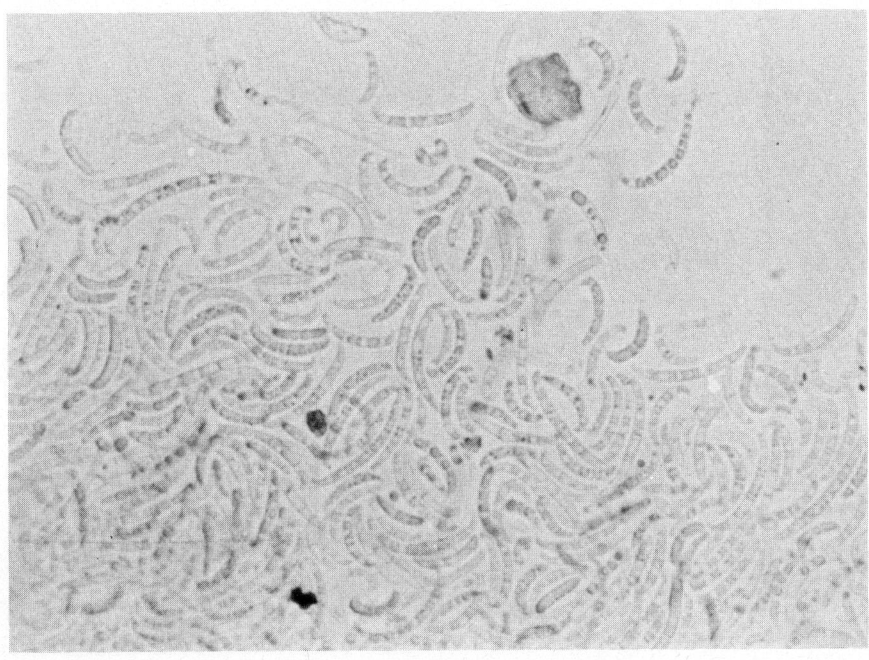

Diagnostic characters : Extremely variable conidia with strongly curved apex and limited host range on *Cajanus* and *Crotalaria*.

(14) FUSARIUM XYLARIOIDES Steyaert, *Bull. Soc. r. Bol. Belg.* **80** : 1-2, 42, 1948

Growth rate 4.0 cm.

Culture pigmentation : pale almost colourless but in some strains becoming deep
violet often with the onset of production of perithecial initials, mycelium
sparse, felted to adpressed.

Conidia variable, curved with uncinate or hooked apex and marked foot cell, 1-septate,
6-10 × 2.5μ, 2-3 septate, 10-30 × 3-3.5μ.

Chlamydospores rare or absent, pre-stroma cells often present.

Perithecial state *Gibberella xylarioides* Heim & Saccas, heterothallic, ascospores
1-3 septate, 15-20 × 5-6.5μ.

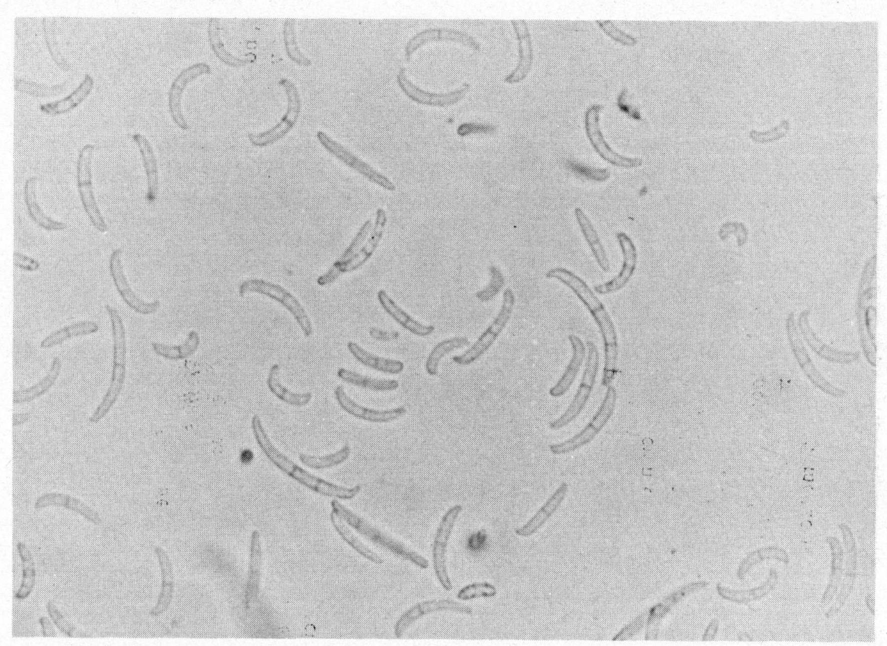

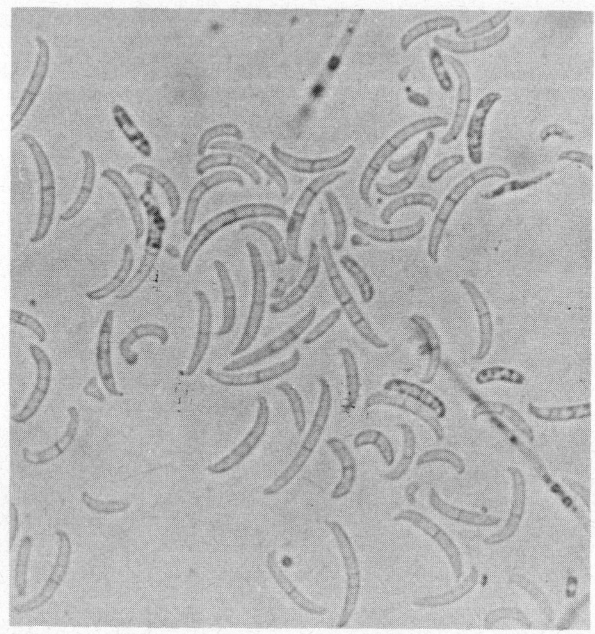

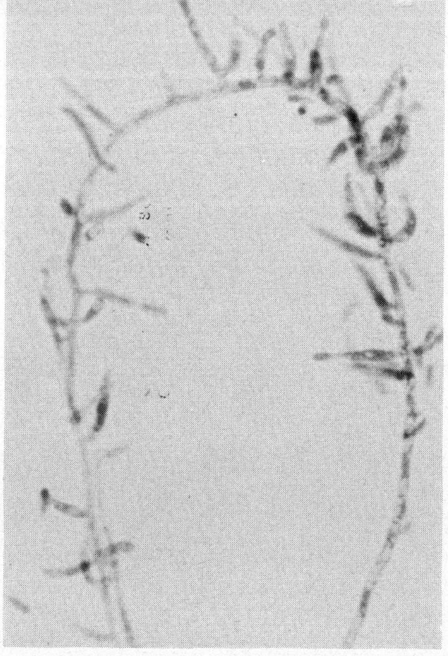

Diagnostic characters : This species causes tracheomycosis and dieback of coffee ;
it is closely related to *F. udum* but differs, apart from host, in pigmentation and in the
presence of a perithecial state.

(15) FUSARIUM FUSARIOIDES (Frag. & Cif.) Booth, *The Genus Fusarium* p. 88, 1971.

Growth rate 4.5 cm.

Culture pigmentation : carmine, coral to red.

Microconidia 0-1 septate, clavate, 8-12 × 2.5-4μ produced as blastospores from apex of sporogenous cell.

Macroconidia, when present, 3-5 septate, 30-46 × 3-5μ, produced from phialides.

Chlamydospores large, globose, up to 30μ diam. and becoming brown, formed in chains or clumps.

Perithecial state unknown.

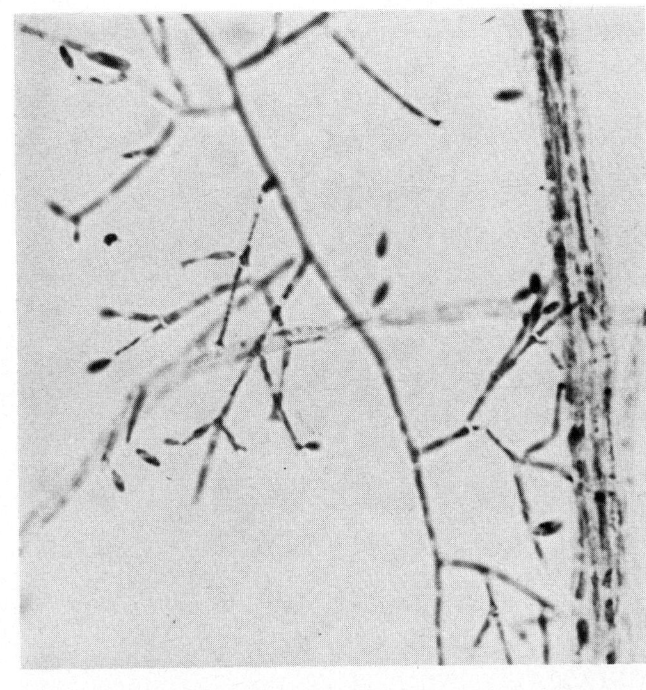

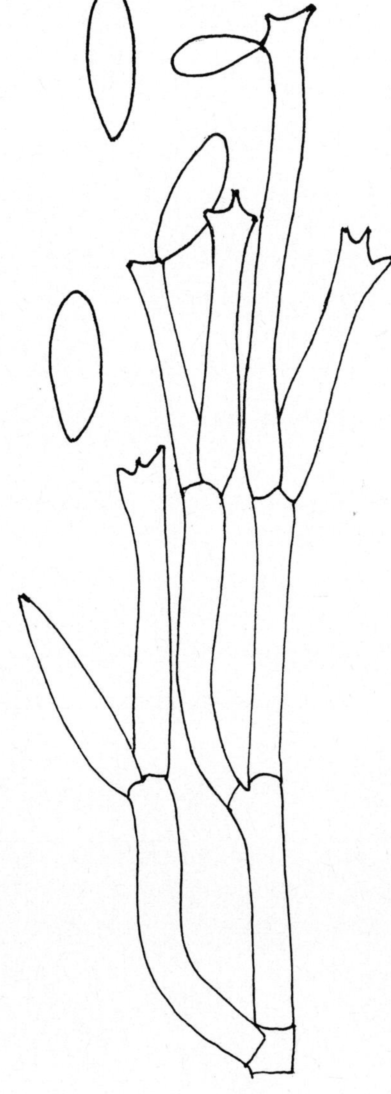

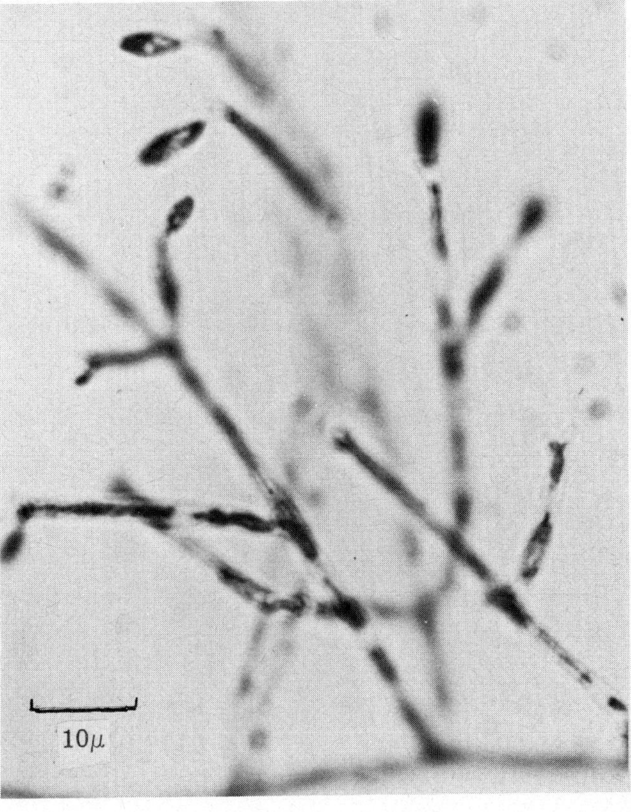

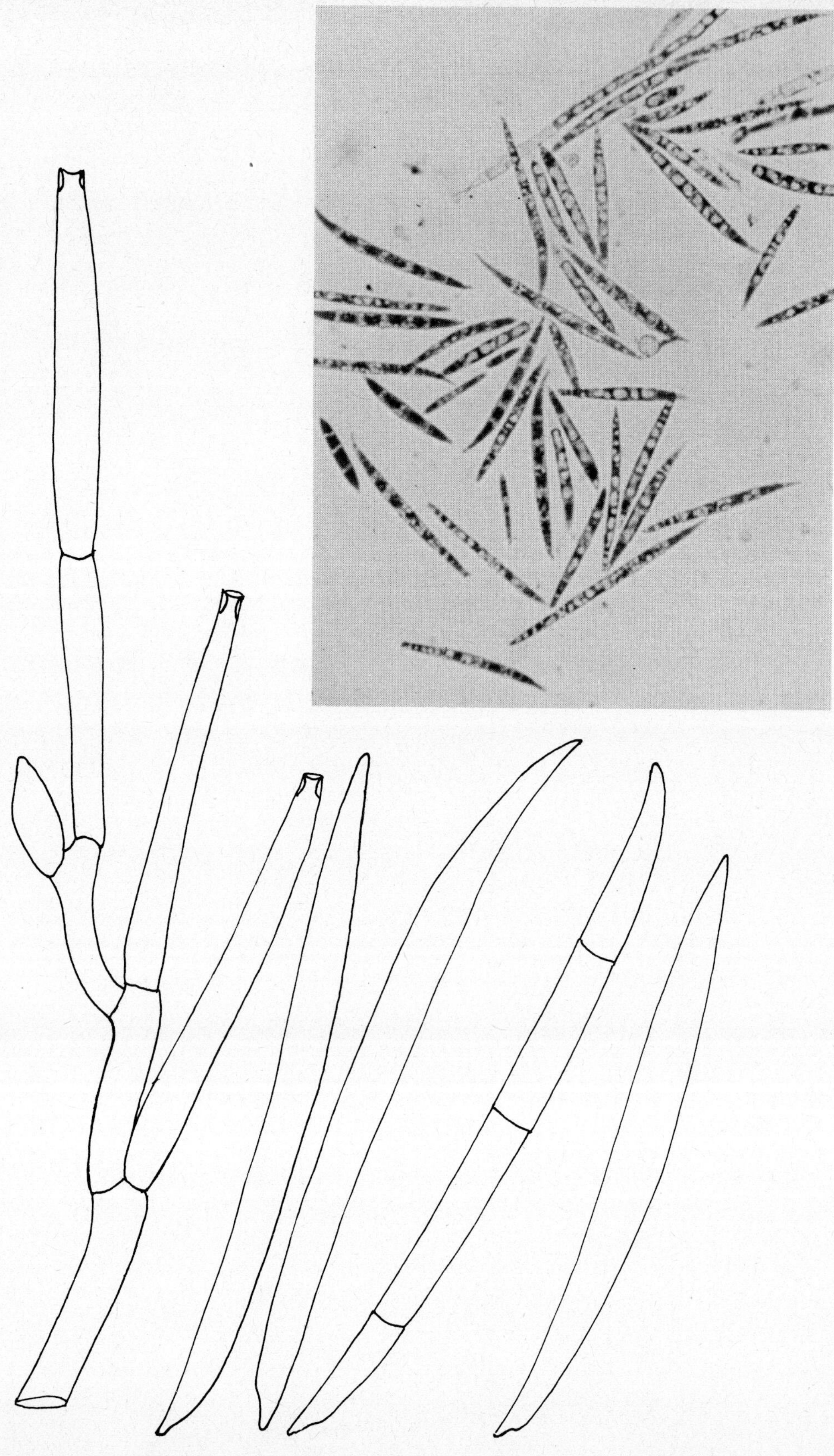

Diagnostic characters : The small clavate microconidia formed as blastospores and the large chlamydospores are characteristic.

(16) FUSARIUM SPOROTRICHIOIDES Sherb., *Mem. Cornell Univ. agric. Exp. Stn.* **6** : 183, 1915.

Growth rate 3.5 cm.

Culture pigmentation : surface mycelium white, agar livid red, becoming tinged with brown.

Microconidia clavate to cymbiform, o-septate, 6-11 × 2.5-4μ, 1-septate, 8-24 × 2.5-4μ, formed as blastospores from pegs more widely dispersed along the sporogenous cell than in *F. fusarioides.*

Macroconidia generally 3-5 septate, 24-50 × 4-5μ.

Chlamydospores, when present, 7-15μ diam., intercalary or in groups.

Perithecial state unknown.

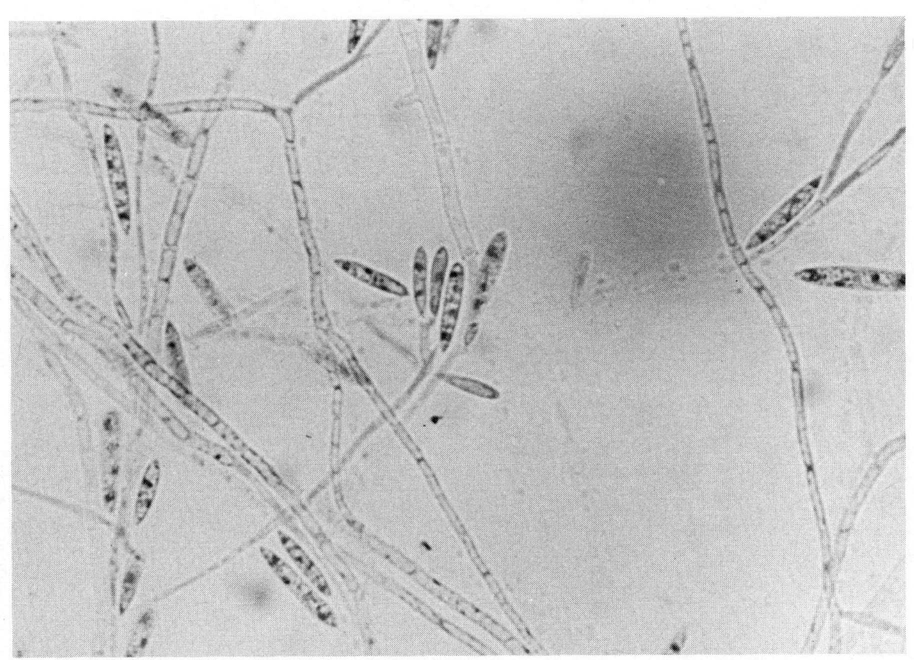

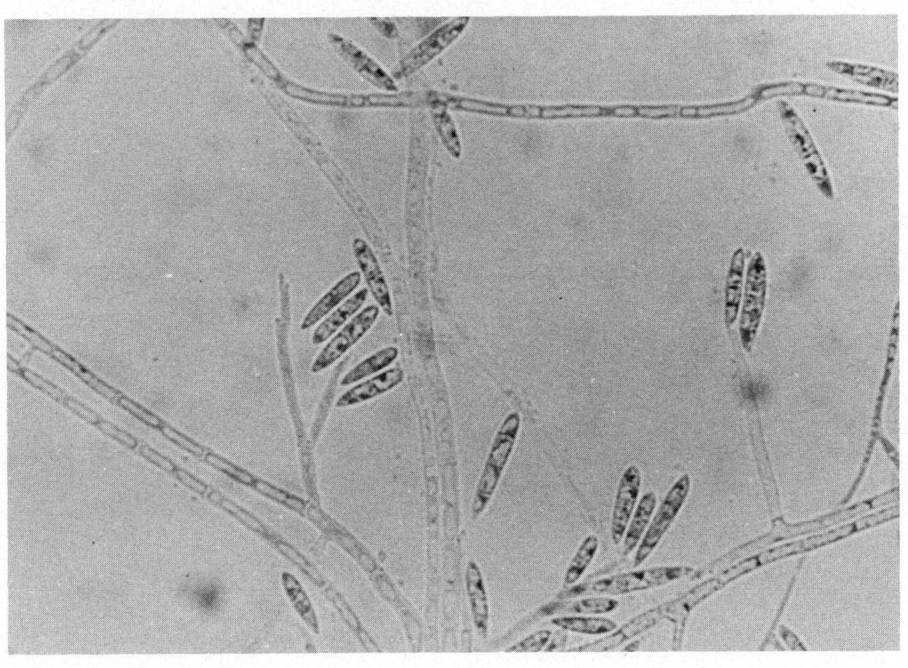

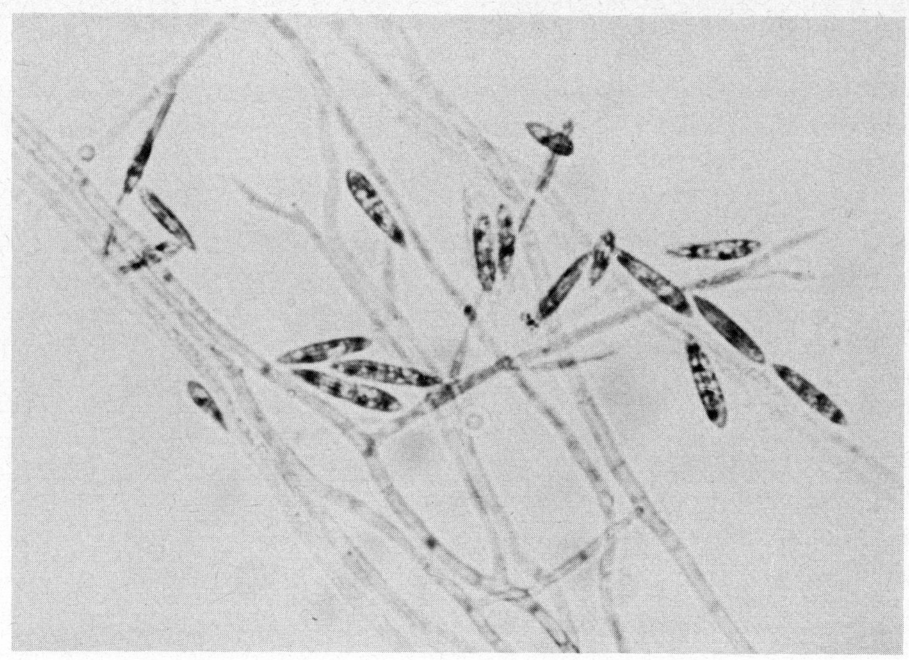

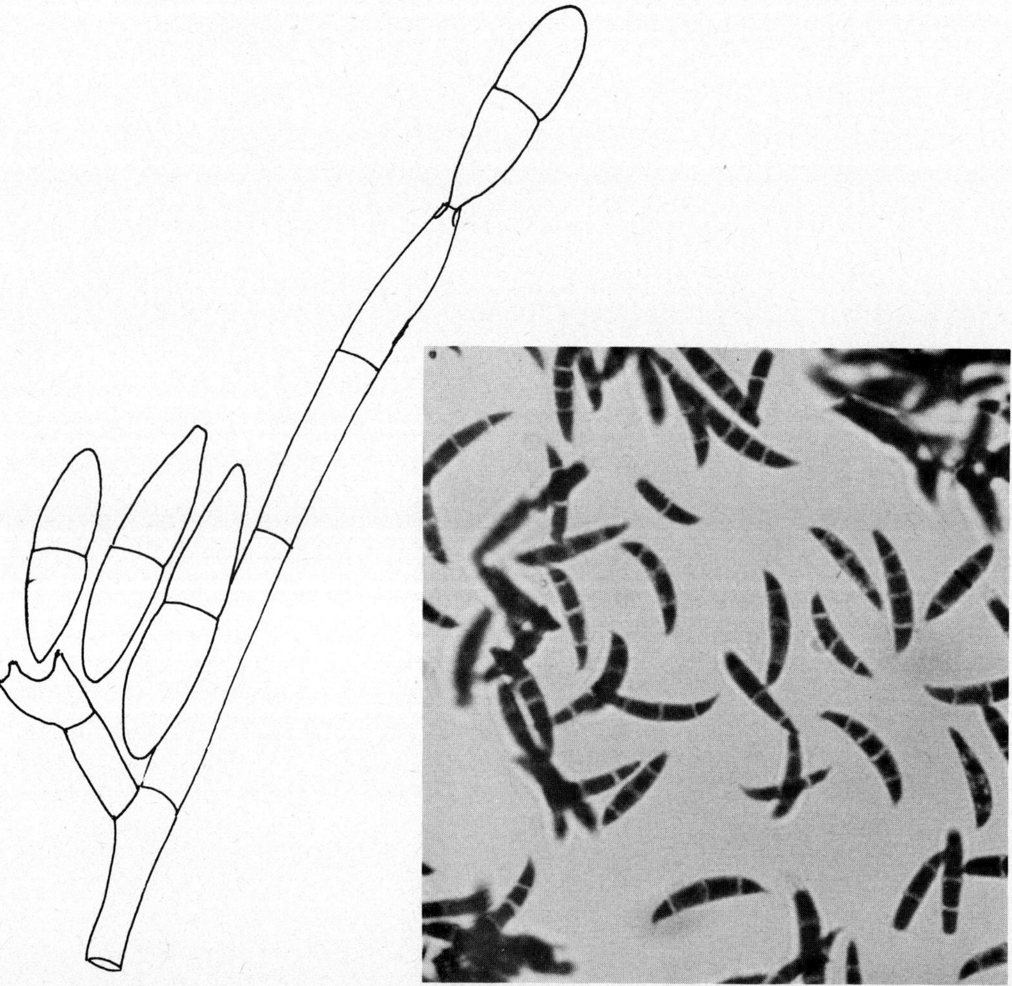

Diagnostic characters : The size, shape and method of formation of the microconidia.

(17) FUSARIUM SEMITECTUM Berk. & Rav. in Berkeley, *Grevillea* **3** : 88, 1875

Growth rate 6.1 cm.

Culture pigmentation peach changing to avellaneous and finally becoming buff brown.

Macroconidia of two types, primary and secondary.

Primary macroconidia with wedge-shape foot cell, 0-5 septate, 7.5-35 × 2.5-4μ, formed as blastospores from polyblastic sympodial cells, up to five separate spores formed by each cell.

Secondary macroconidia with typical heeled foot-cell, 3-7 septate, 20-46 × 3-5.5μ, formed from phialides usually grouped in sporodochia.

Chlamydospores often sparse, globose, 10-12μ diam., becoming brown, intercalary, single or in chains.

Perithecial state not reported.

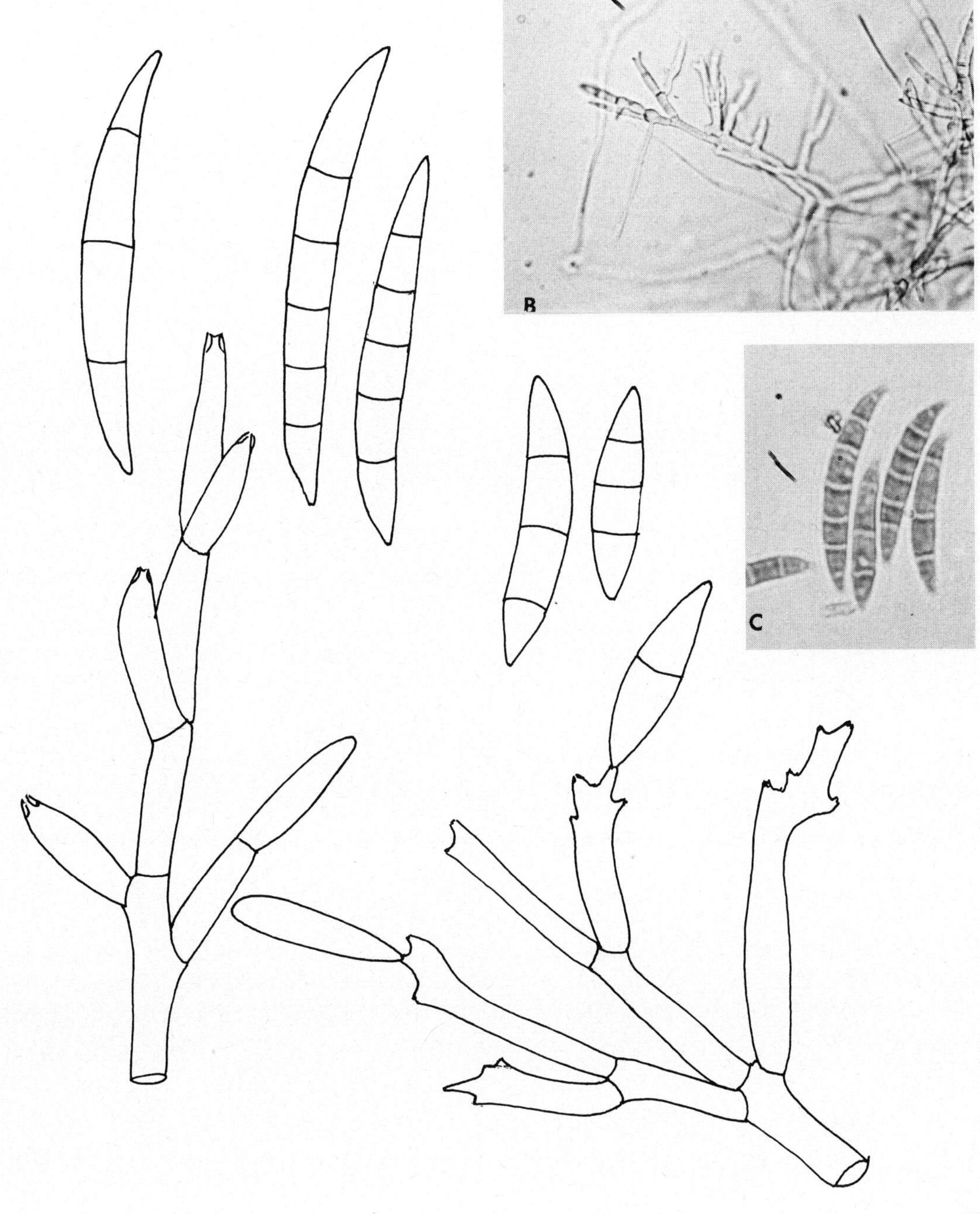

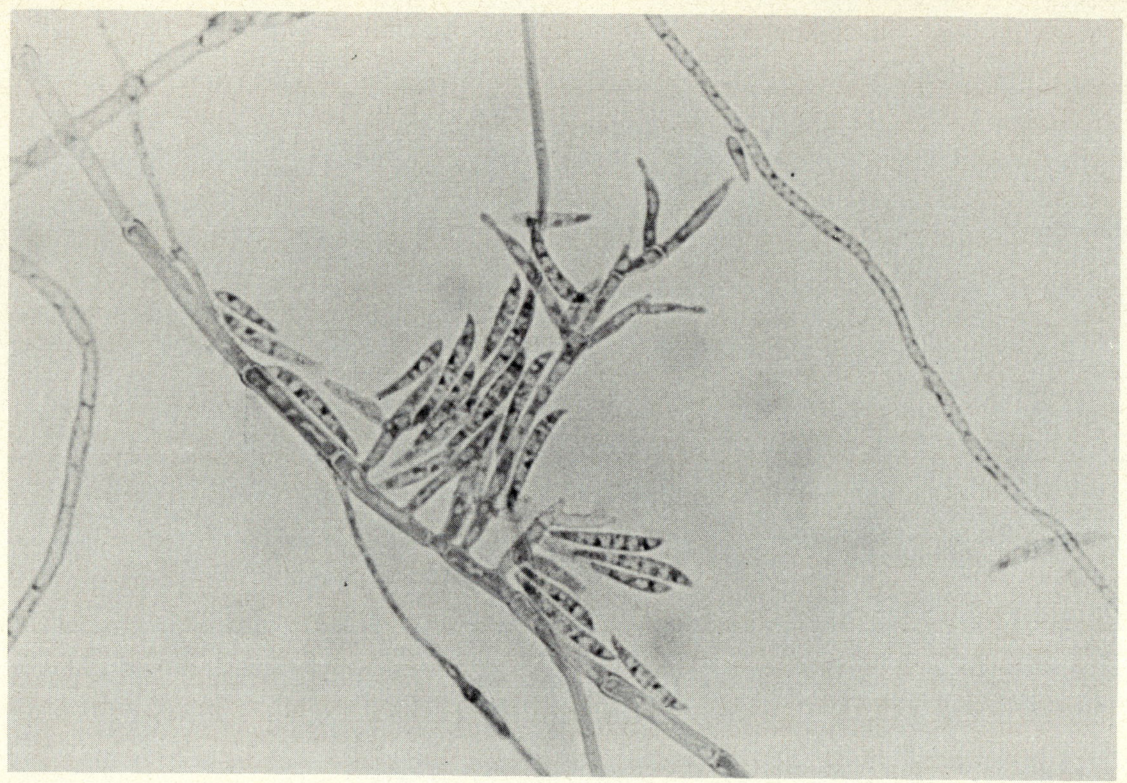

Diagnostic characters : Both *F. semitectum* and *F. avenaceum* produce primary (blastosporic) and secondary (phialidic) macroconidia in fresh isolates. They are clearly separated on pigmentation and spore form and by the presence of chlamydospores in *F. semitectum*.

(18) FUSARIUM AVENACEUM (Fr.) Sacc., *Sylloge Fung.* **4** : 713, 1886.

Growth rate 5.4 cm.

Culture pigmentation : aerial mycelium rose red fringed with white, yellowish brown from below.

Macroconidia of two types, primary and secondary, in fresh isolates.

Primary macroconidia fusoid, 1-3 septate, 8-50 × 3.5-4.5μ produced from polyblastic conidiogenous cells.

Secondary macroconidia are the typical form produced from phialides often developing in sporodochia ; conidia 4-7 septate, 40-80 × 3.5-4μ.

Chlamydospores absent in mycelium, rarely formed in conidia.

Perithecial state *Gibberella avenacea* Cooke; not confirmed.

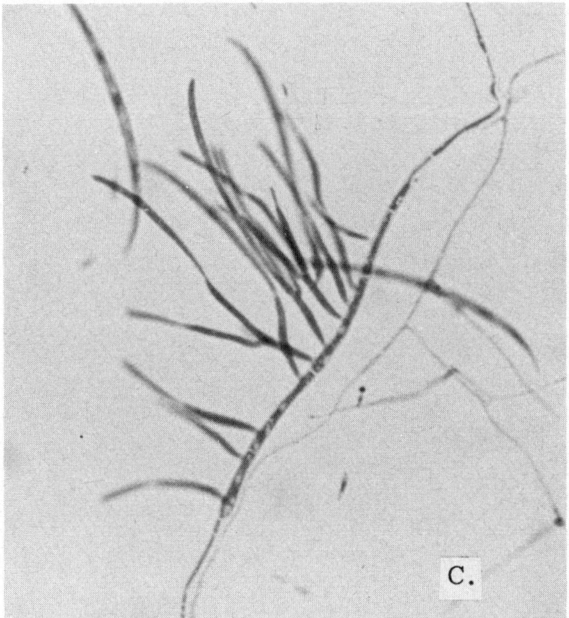

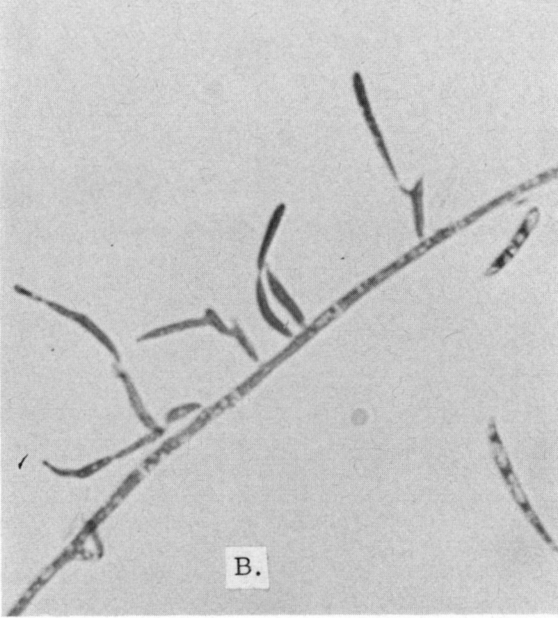

A, mature secondary conidia.
B, formation of primary conidia,
C, secondary conidia.

Diagnostic characters : Presence of primary (blastosporic) and secondary (phialidic) macroconidia. The long narrow secondary macroconidia, absence of chlamydospores and culture pigmentation are also useful points of separation.

(19) FUSARIUM GRAMINEARUM Schwabe, *Fl. Anhaltina* **2** : 285, 1838

Growth rate 8.9 cm.

Cultural pigmentation : rose, coral becoming vinaceous with a brown tinge.

Macroconidia only produced from simple lateral phialides which may or may not become grouped on branched conidiophores. Macroconidia falcate generally with an elongated apical cell narrowing gradually to a point ; 3 septate, 30-50 × 3.5-4μ, 5-7 septate, 36 × 3.5-5μ.

Chlamydospores absent or rare ; if present intercalary, 10-12μ diam.

Chromosome number = 6.

Perithecial state *Gibberella zeae* Schwabe; ascospores 3 septate, 20-24 × 4-5μ.

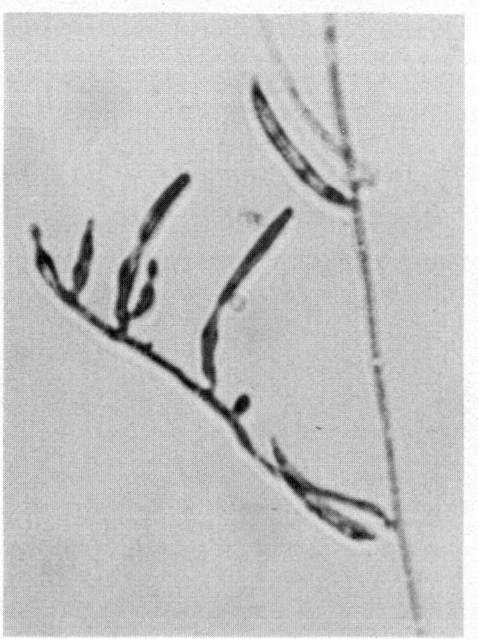
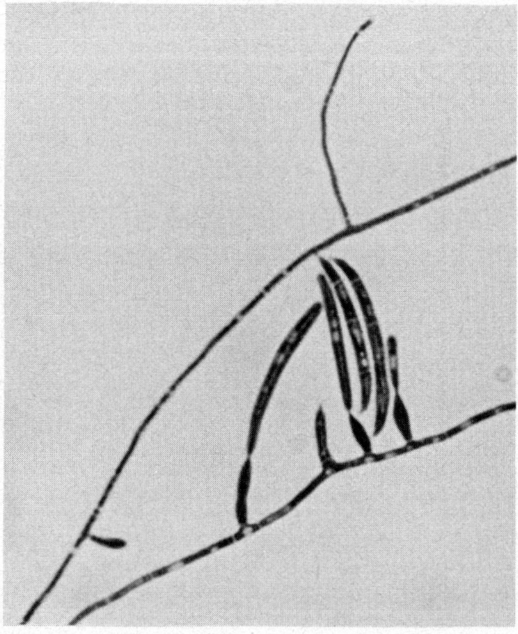
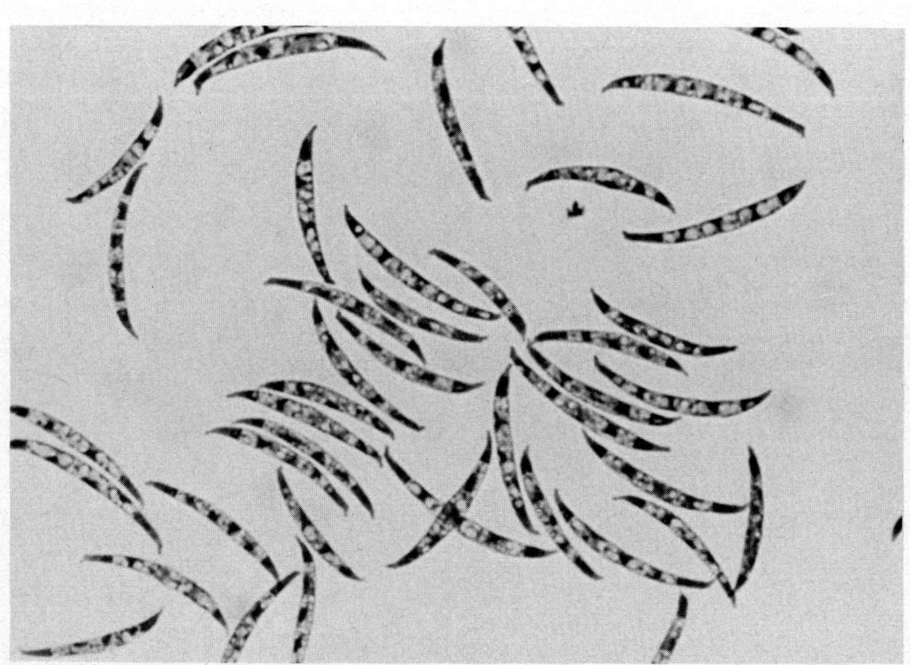

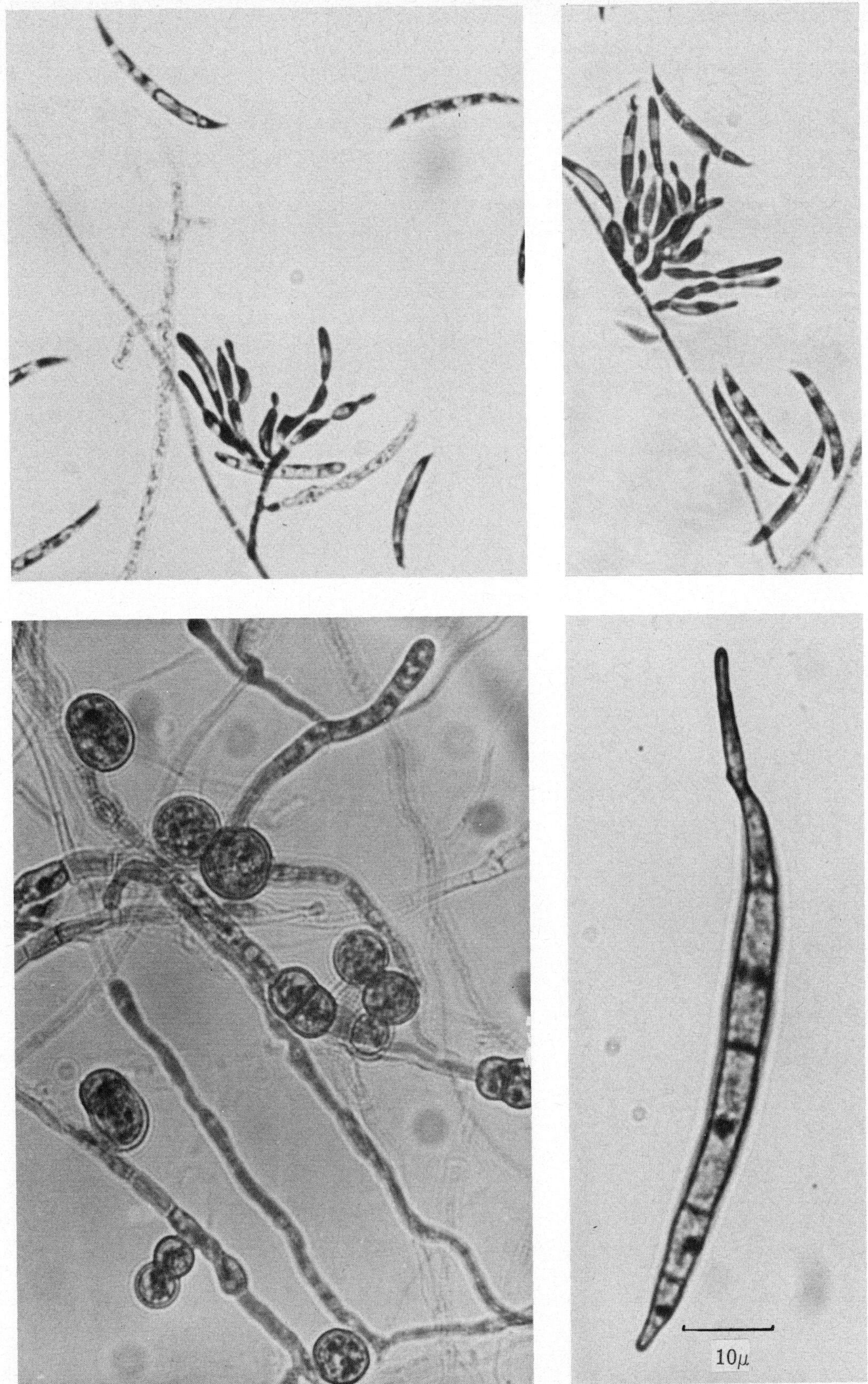

Diagnostic characters : The long falcate macroconidia often formed sparsely in many strains are characteristic. Many isolates of this species with floccose aerial mycelium and rose to coral pigmentation produce neither macroconidia nor chlamydospores until surface of colony is washed clean of mycelium and culture reincubated.

(20) FUSARIUM HETEROSPORUM Nees ex Fr., *Syst. mycol.* 3 : 472, 1832.

Growth rate = 4.2 cm.

Culture pigmentation pale pink to delicate peach, conidial sporodochia deep orange. Some strains show reddish pigmentation later.

Macroconidia only, produced from simple phialides, are curved with an elongated apical cell ; 3 septate, 17-40 $\times$ 3-3.5μ, 5 septate, 38-55 $\times$ 4μ.

Chlamydospore formation very sparse.

Perithecial state *Gibberella gordonia* Booth ; ascospores 1-3 septate, 15.6-18.5 $\times$ 4-4.5μ.

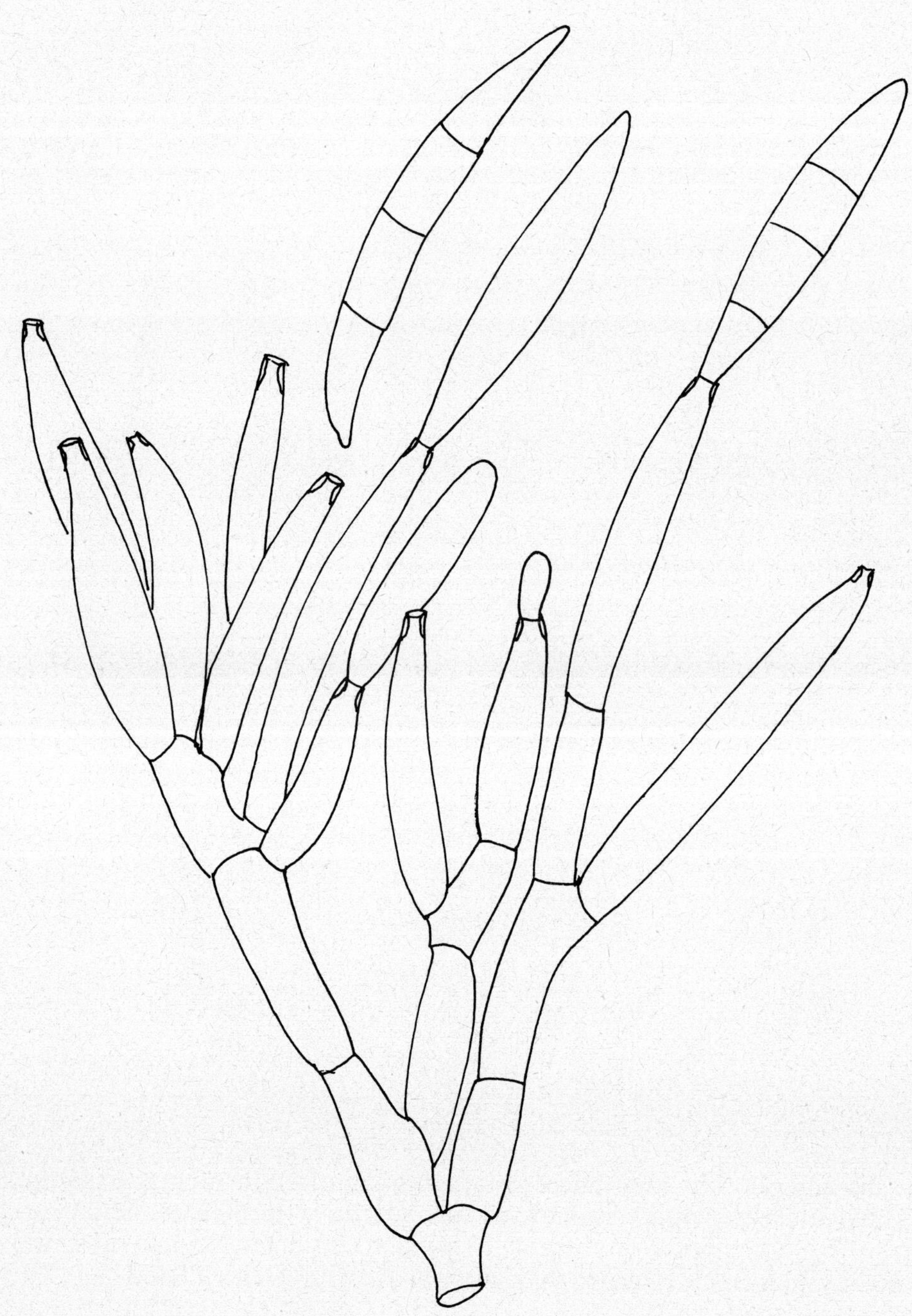

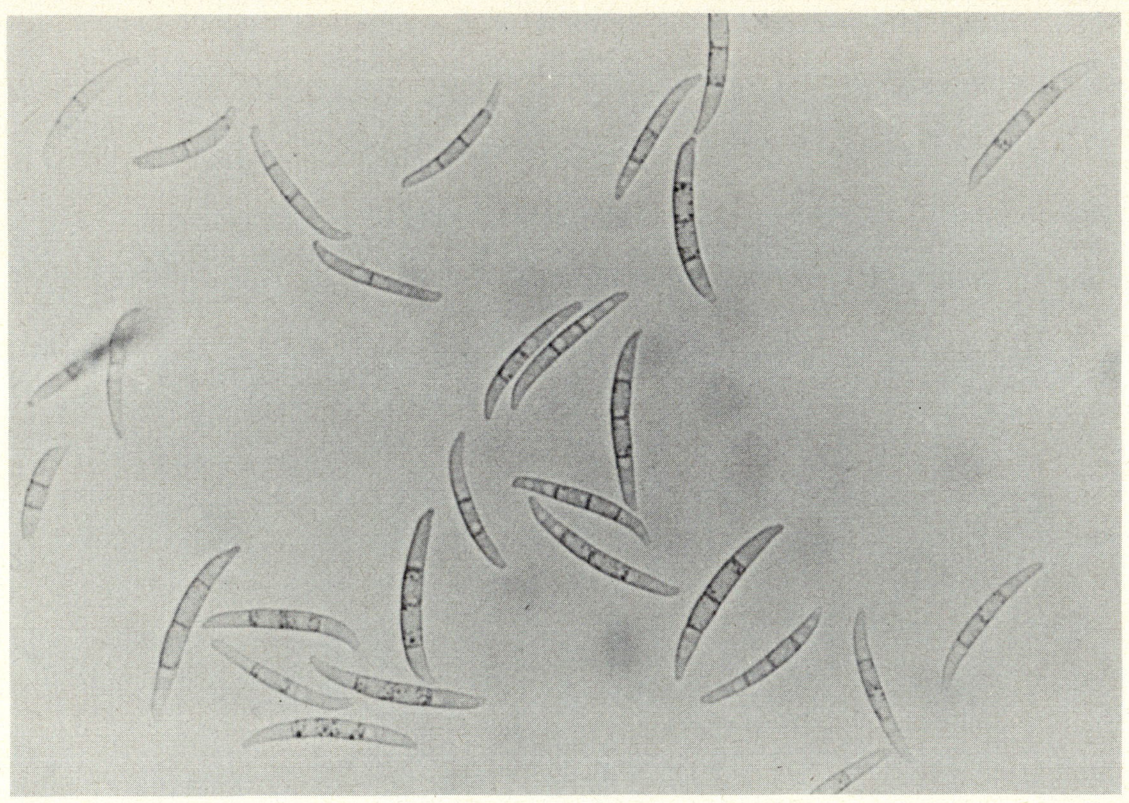

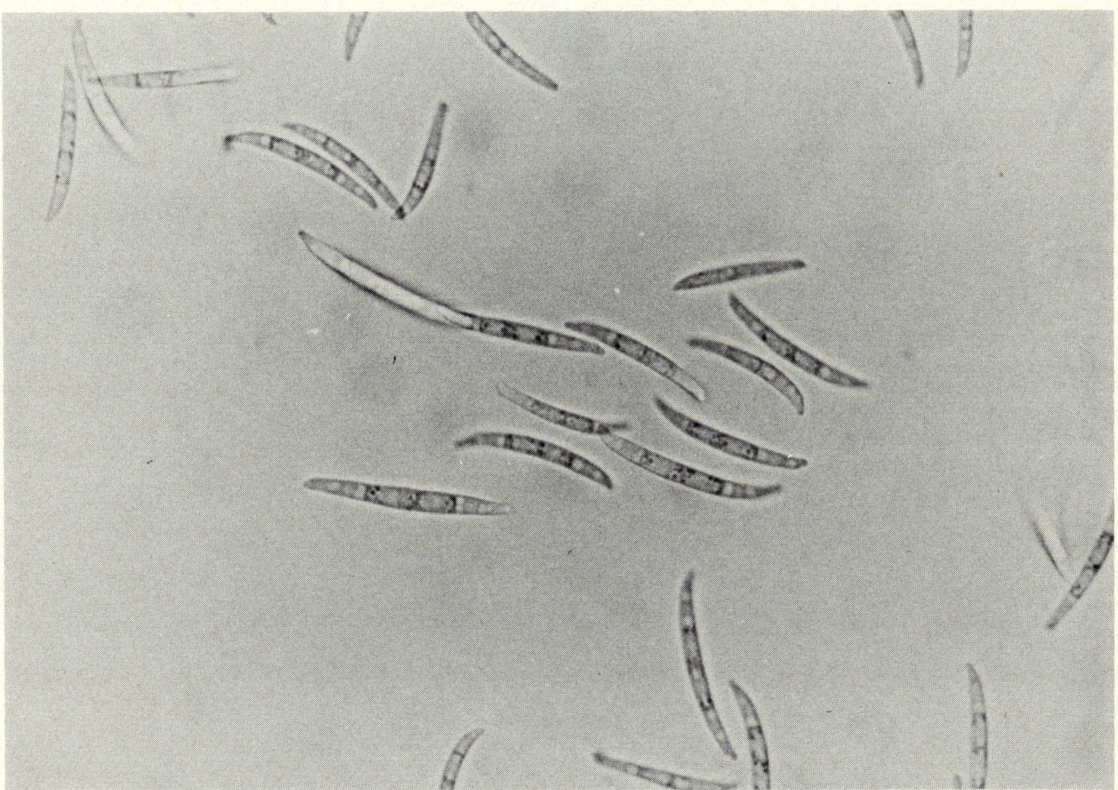

Diagnostic characters : This is the only *Fusarium* species whose macroconidial production shows marked response to irradiation with near UV light. The name was first used for the *Fusarium* with the above characters which causes head blight of cereals and its use should be confined to these strains.

(21) FUSARIUM SULPHUREUM Schlecht., *Flora berol.* p. 139, 1824.

Growth rate 6.4 cm.

Cultural pigmentation : cream, sulphureus to light brown.

Macroconidia only produced, 3 septate, 23-32 $\times$ 3.5-4μ, 5 septate, 30-46 $\times$ 4.5μ.

Chlamydospores globose 8-10μ diam., formed sparsely, singly or in chains.

Perithecial state *Gibberella cyanogena* (Desm.) Sacc., ascospores 3 septate, 20-25 $\times$ 5.7μ.

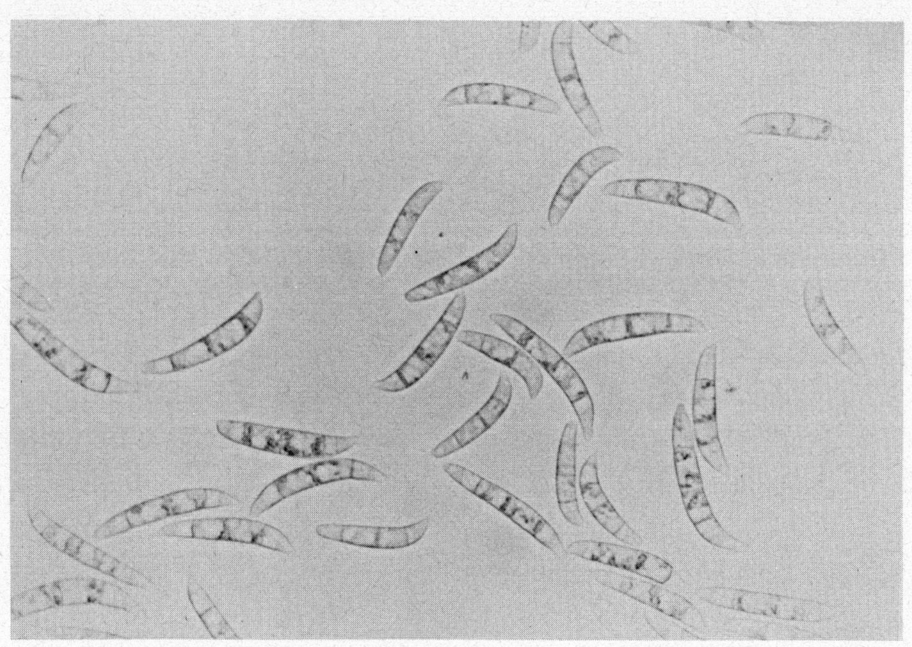

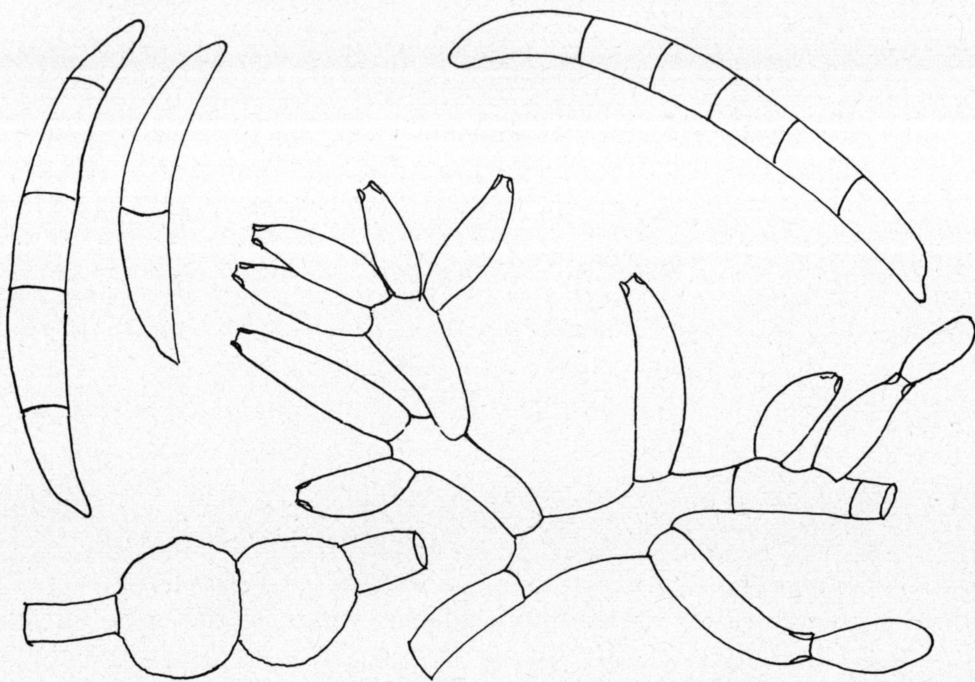

Diagnostic characters : This species is allied to *F. sambucinum* but the culture pigment is distinct and the apical cells of the macroconidia are not usually beaked ; ascospores are also narrower.

44

(22) FUSARIUM SAMBUCINUM Fuckel, *Symb. mycol.* p. 167, 1869.

Growth rate 5.2 cm.

Culture pigmentation : peach to orange or ochreous, in some isolates vinaceous to bay.

Macroconidia only produced ; strongly dorsiventral with beaked apical cell ; 3 septate, 25-40 × 4-5.5μ, 5 septate, 40-53 × 4-5.5μ.

Chlamydospores, formed sparsely in knots or chains, are globose, 6-11μ diam.

Perithecial state *Gibberella pulicaris* (Fries) Sacc.; ascospores 3 septate, 20-28 × 6-9μ.

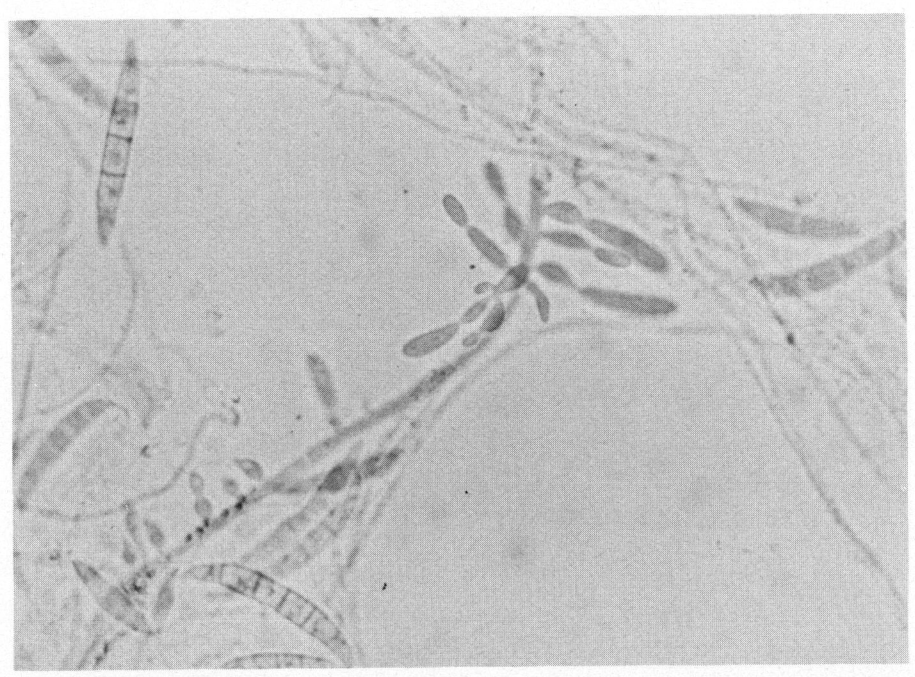

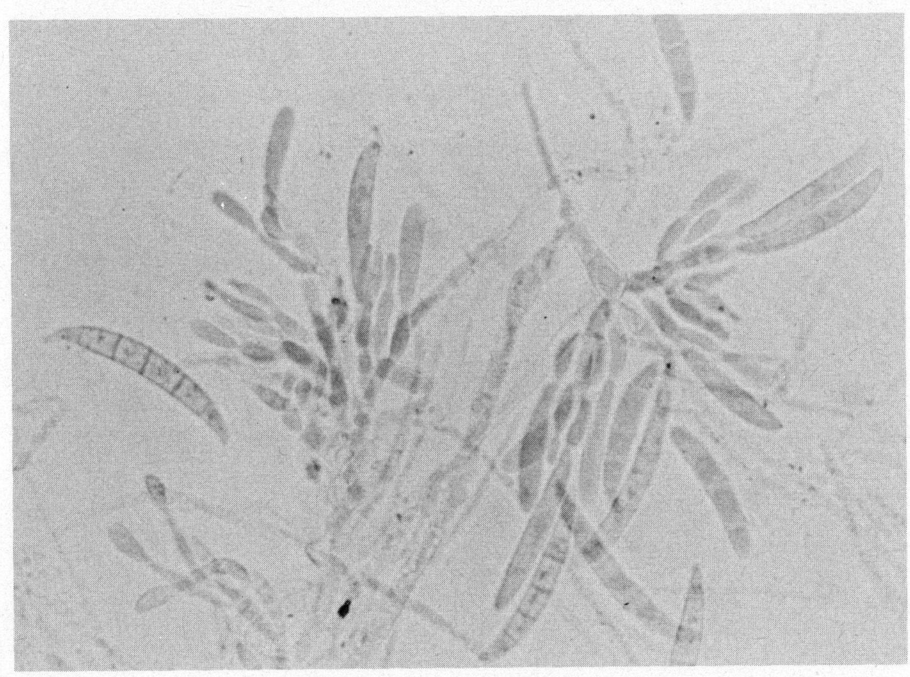

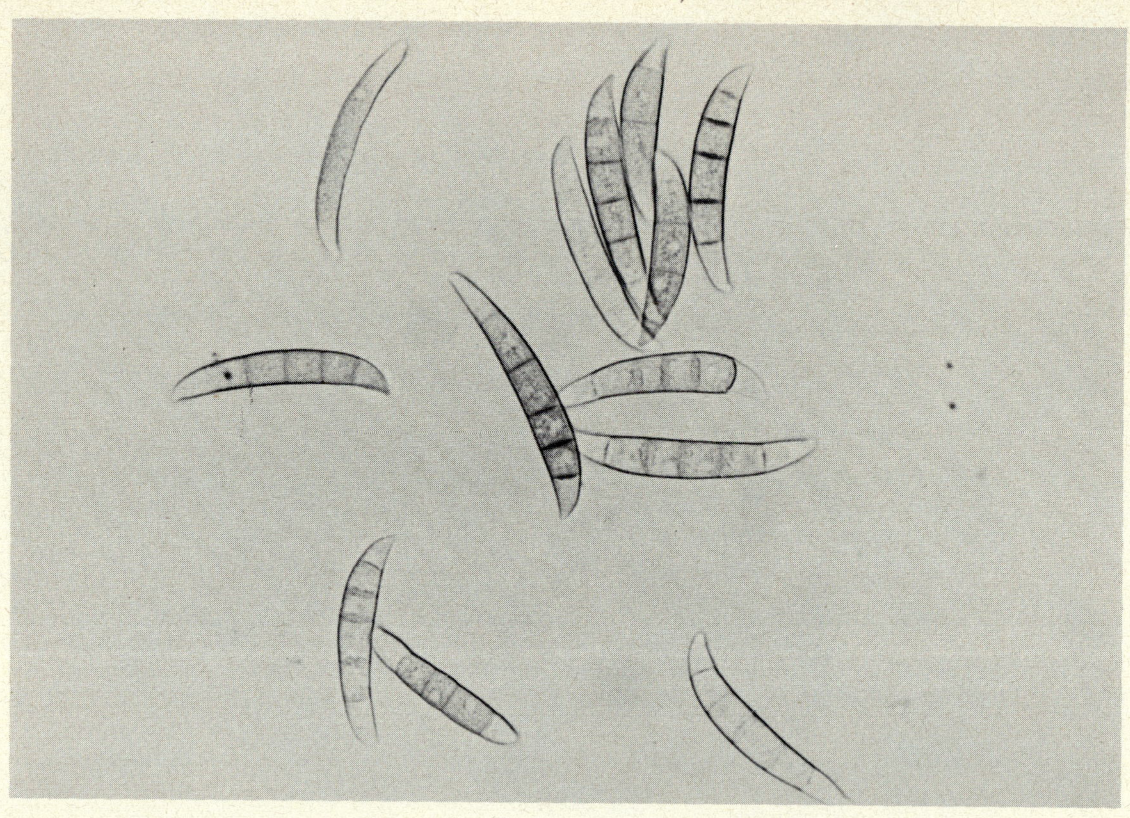

Diagnostic characters : Macroconidia resemble those of *F. culmorum* in shape ; they are approximately the same length but are narrower. Cultures also have a slower growth rate and a well known perithecial state.

(23) FUSARIUM SAMBUCINUM Fuckel var. COERULEUM Wollenw., *Annls mycol.* **15** : 55, 1917

Growth rate 4.8 cm.

Cultural pigmentation : olivaceous buff to olivaceous or cinnamon and in other strains reddish-purple.

Macroconidia only produced. They are strongly curved and with a beaked apical cell ; 3 septate, 17-30 × 5-6μ, 5 septate, 25-37 × 4.5-6μ.

Chlamydospores intercalary, singly or in chains, 10-14μ diam.

Perithecial state unknown.

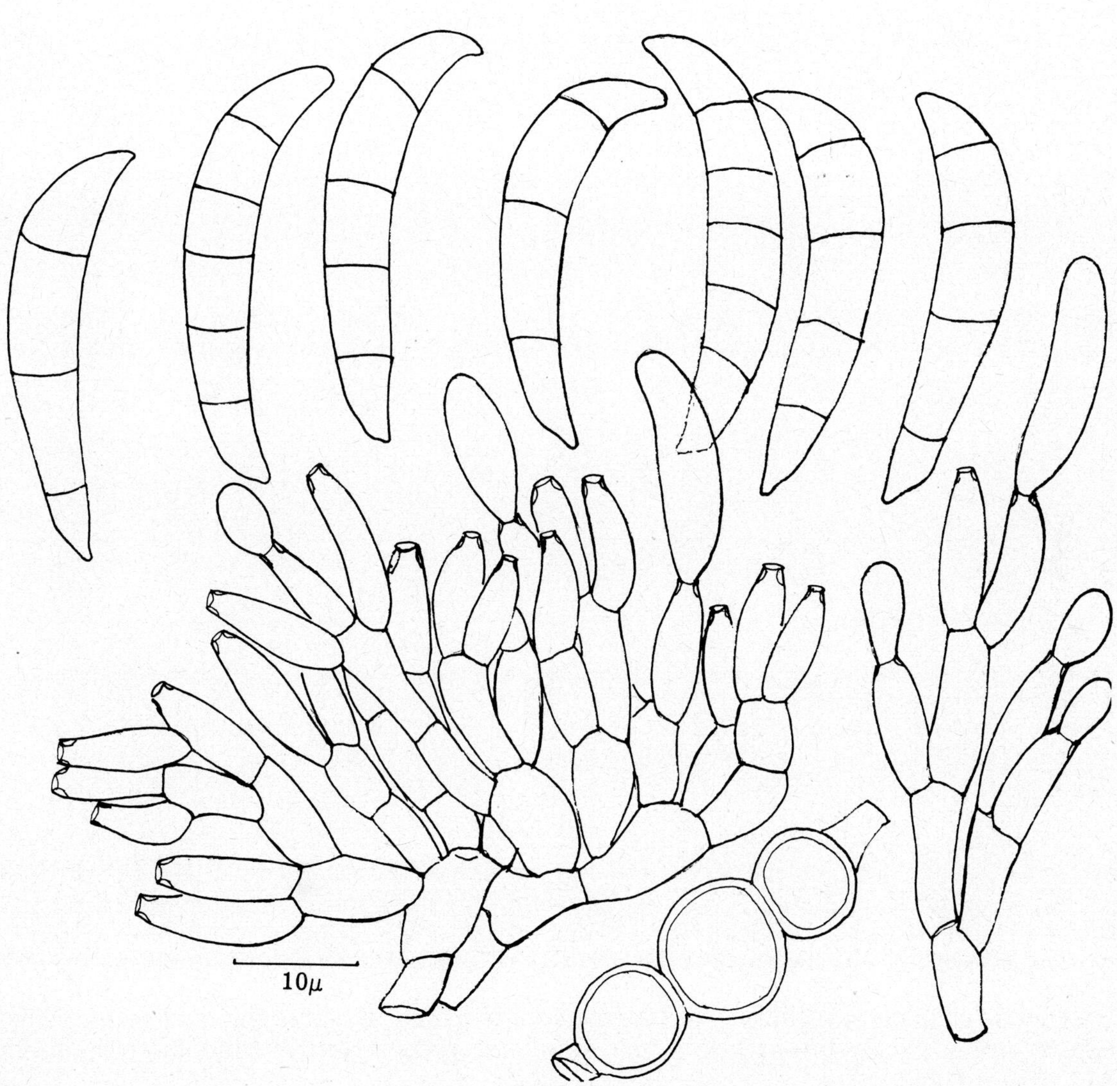

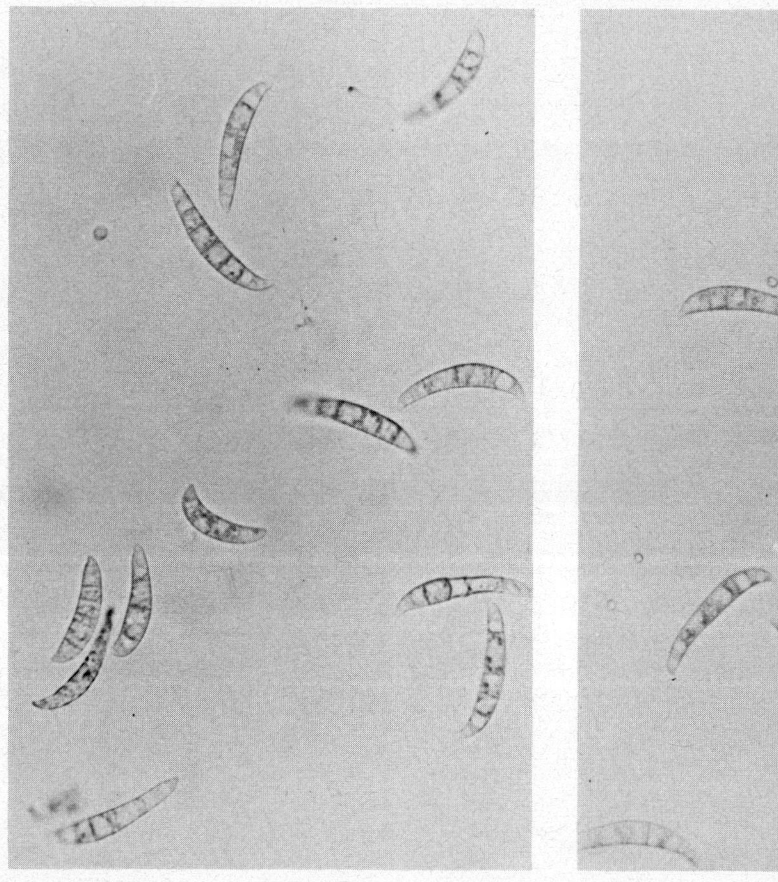

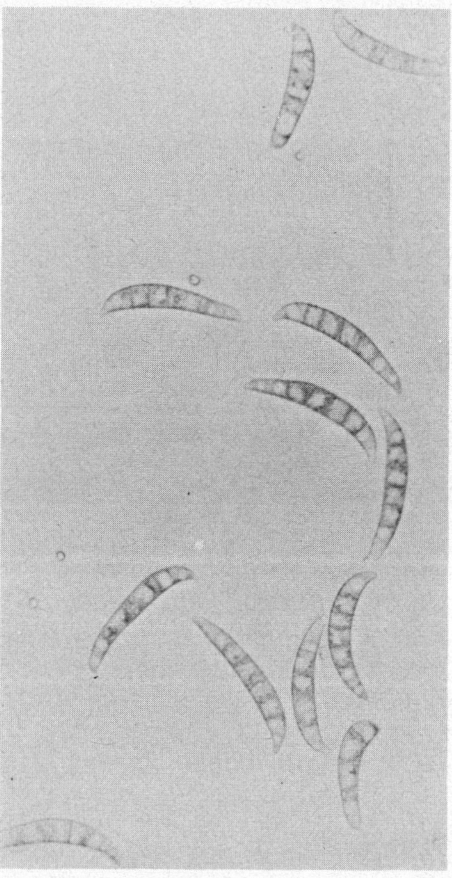

Diagnostic characters : The shorter curved spores readily separate this variety from the parent strain. Spore production when abundant gives greenish-blue discolouration on surface of culture.

(24) FUSARIUM CULMORUM (W. G. Smith) Sacc., *Sylloge Fung.* **II** : 651, 1885.

Growth rate 8.5 cm.

Cultural pigmentation : red becoming reddish-brown.

Macroconidia only produced ; very uniform in shape, 3 septate, 27-36 $\times$ 5-7μ, 5 septate, 31-50 $\times$ 5-7.5μ. They are produced from simple phialides borne on complex, loosely branched conidiophores.

Chlamydospores oval to globose, smooth to rough walled, 10-14 $\times$ 9-12μ formed singly, in chains or clumps.

Chromosome number = 8.

Perithecial state unknown.

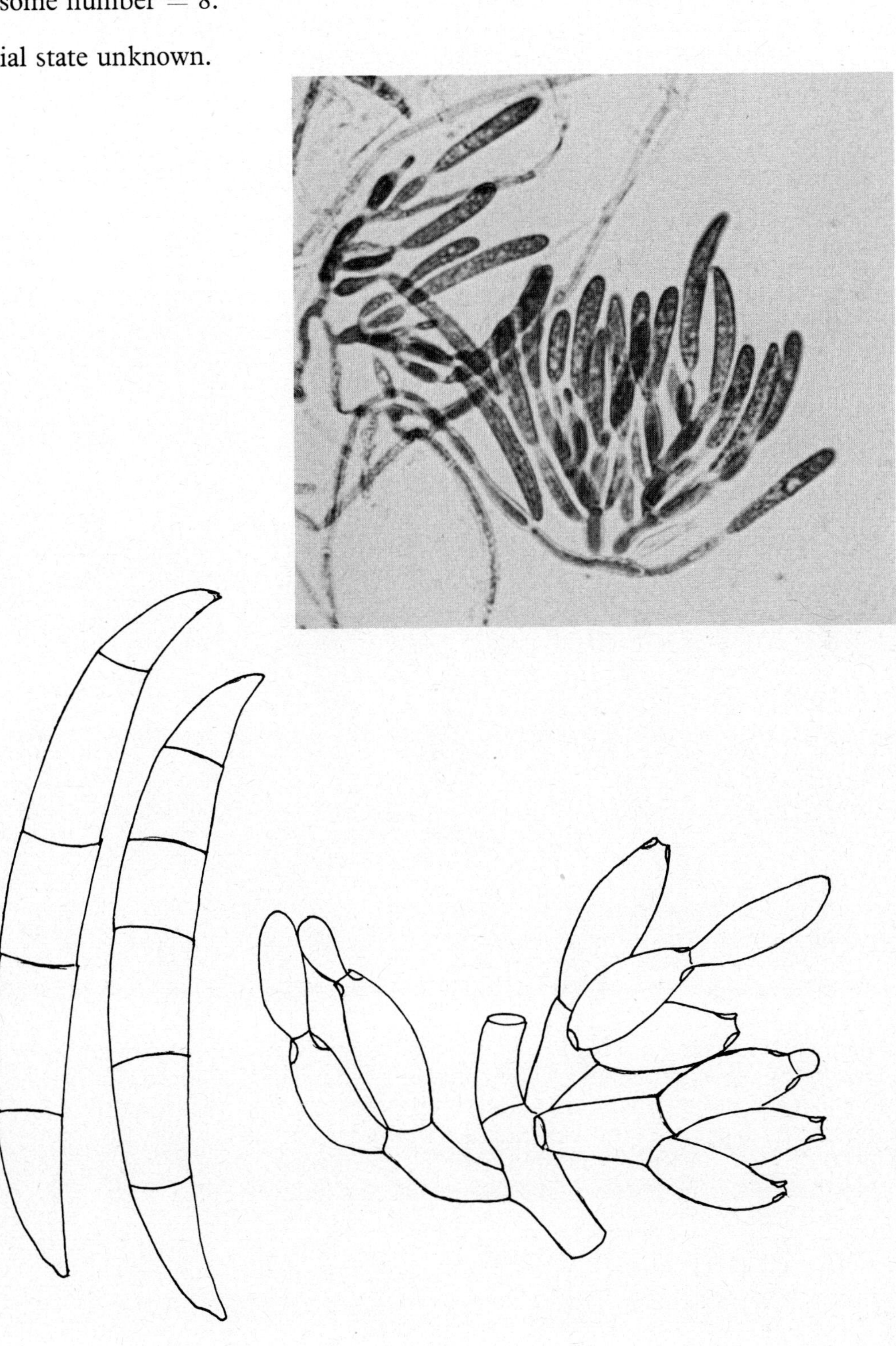

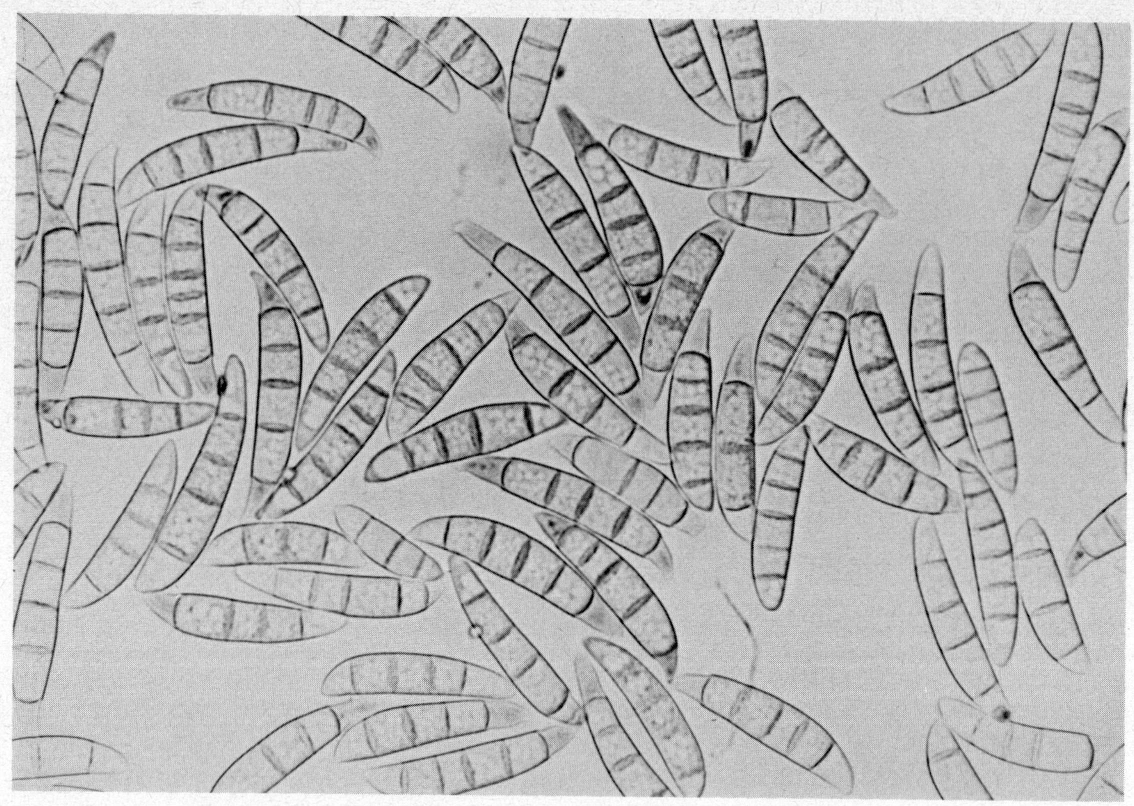

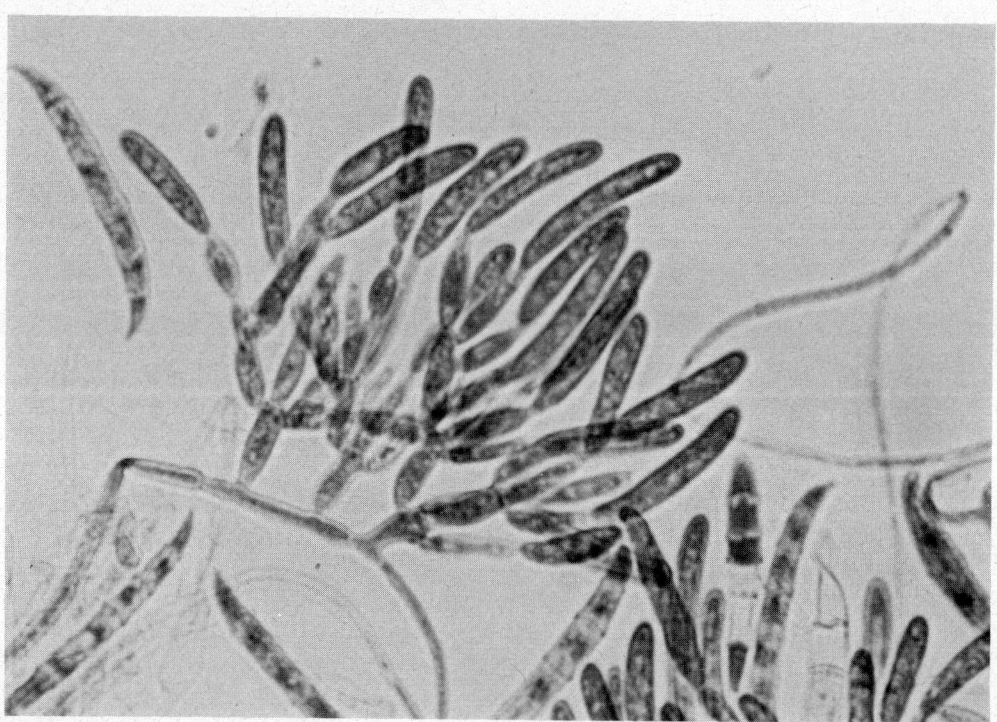

Diagnostic characters : This is a very stable species readily separated from *F. sambucinum* by the wider conidia. The conidial width, shape of apical cell in, and regular presence of, macroconidia and chlamydospores separate it from *F. graminearum*. Two other diagnostic features are a : the secretion of an oily non-water soluble substance in 2-3 wks old cultures and b : the *in situ* germination of macroconidia to produce aberrant microconidia from small phialides.

(25) FUSARIUM NIVALE (Fr.) Ces., *Rabenh. Klotzsch, Herb. Mycol.* Ed. 1. No. 1439, 1850.

Growth rate 1.3 cm.

Culture pigmentation : white to pale peach, occasionally apricot colour on the surface.

Conidia broadly falcate, 1-3 septate, 10-30 × 2.5-5μ.

Chlamydospores not observed.

Perithecial state *Micronectriella nivalis* (Schaffn.) Booth, homothallic, ascospores 1 occasionally 2-3 septate, 10-17 × 3.5-4.5μ.

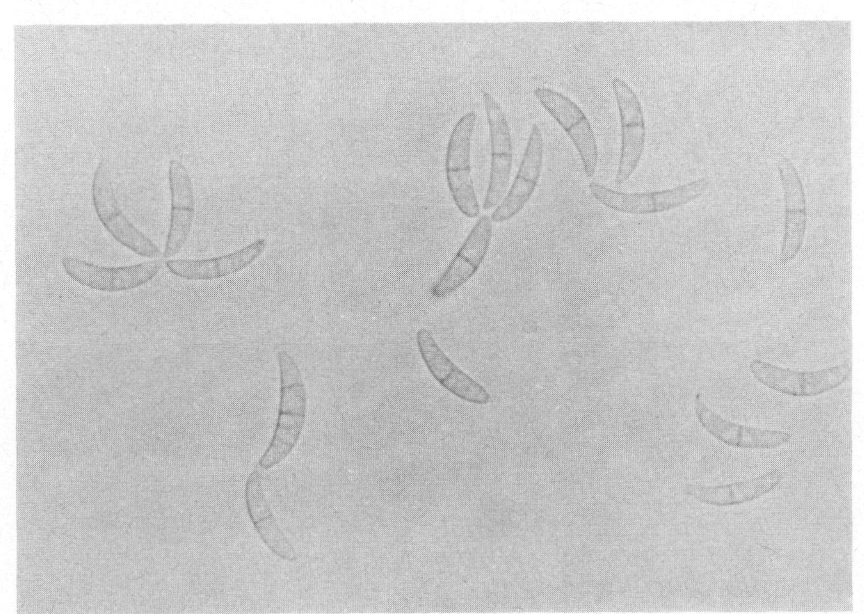

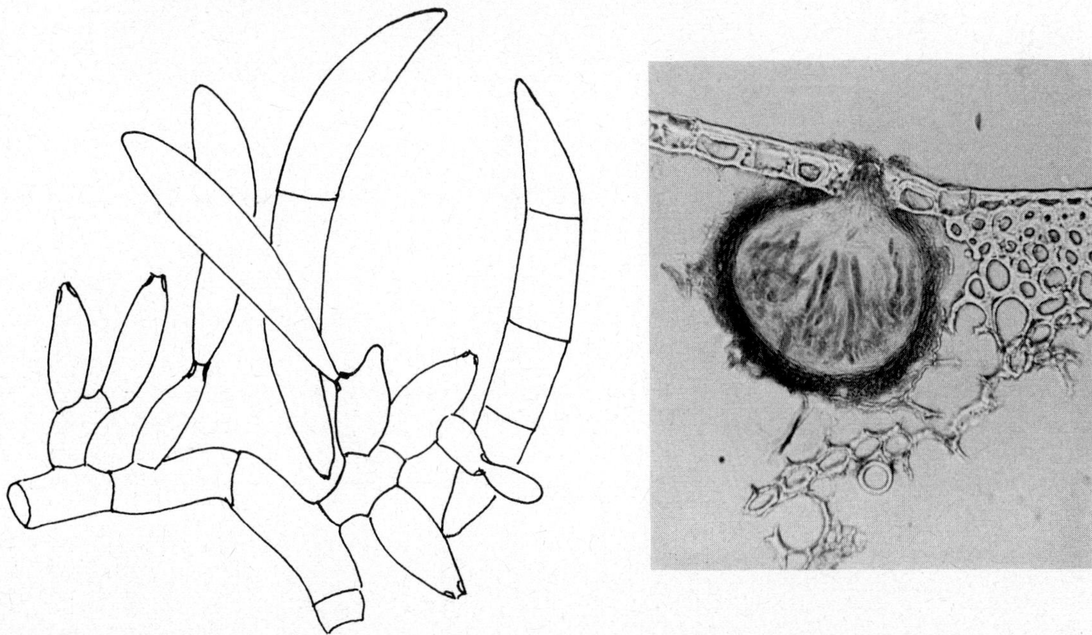

Diagnostic characters : Slow growth, culture pigmentation, spore shape and size and the presence of abundant proliferating phialides. Perithecial state readily develops in fresh isolates.

(26) FUSARIUM DIMERUM Penz. in Saccardo *Michelia* **2** : 484, 1882.

Growth rate 2.7 cm.

Cultural pigmentation : orange beige to apricot.

Conidia somewhat heterogenous probably representing primary and secondary conidia as occasional polyphialides develop, 0-septate, 6.5-10.5 × 2.3-2.5μ, 1-2 septate, 10-22 × 3-3.5μ.

Chlamydospores globose, oval to smooth, 8-12μ diam., intercalary, formed, singly or in chains.

Perithecial state unknown.

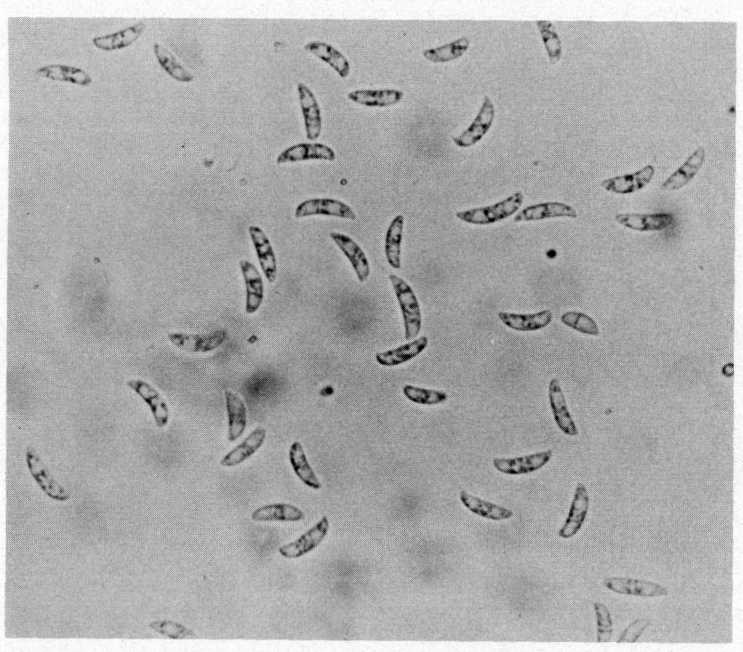

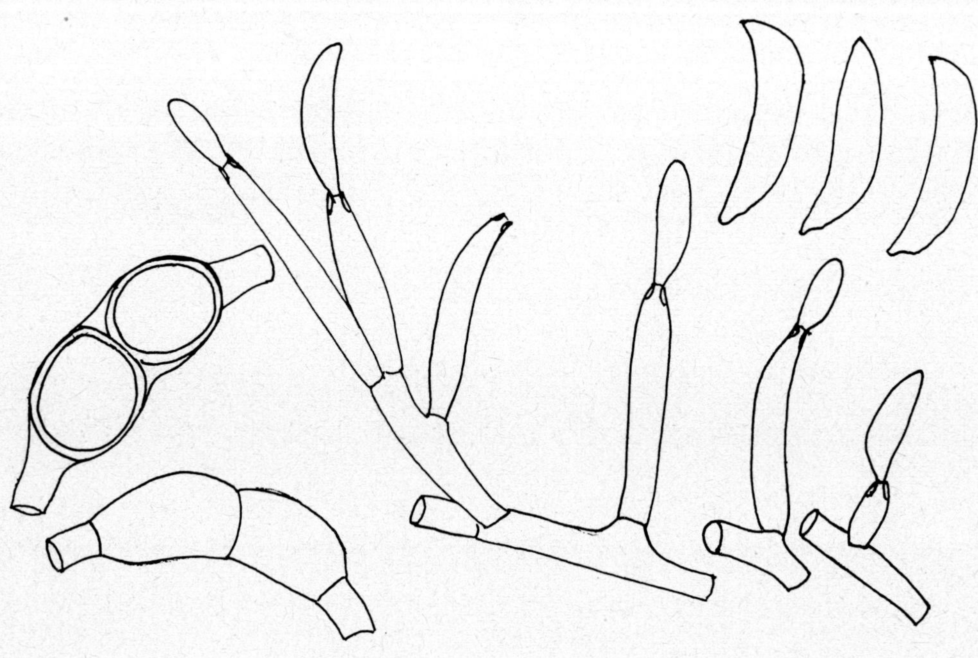

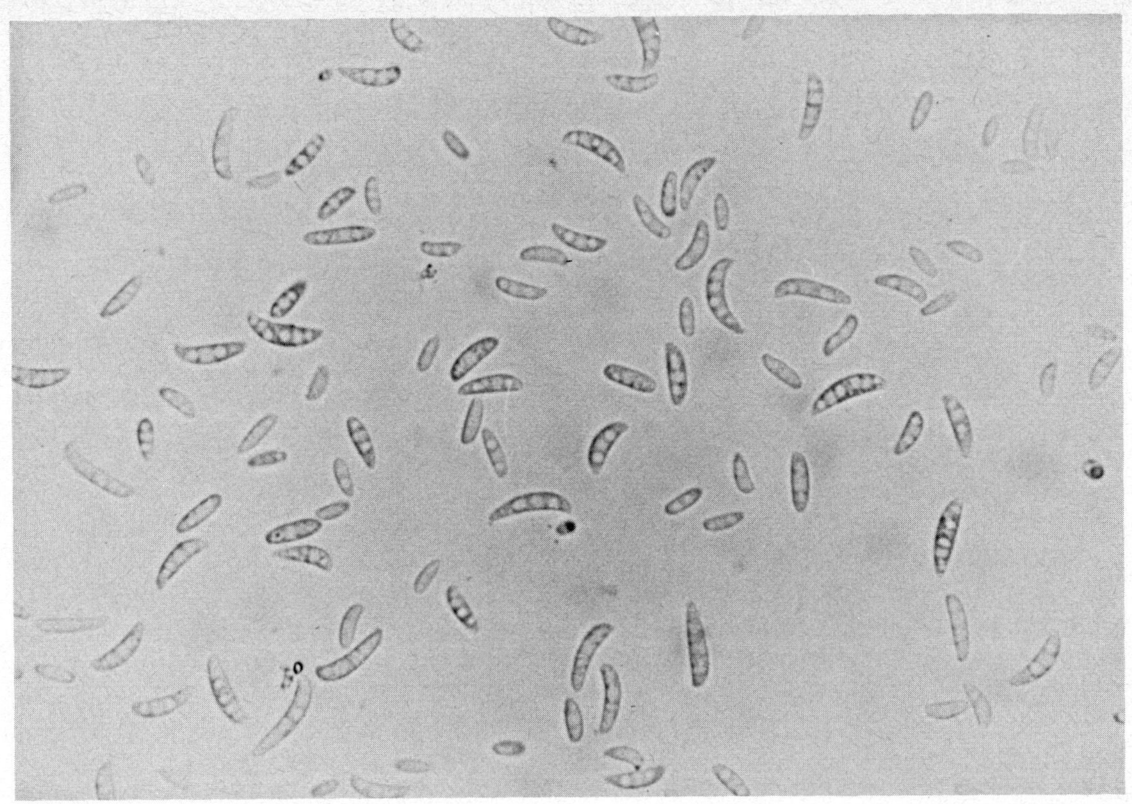

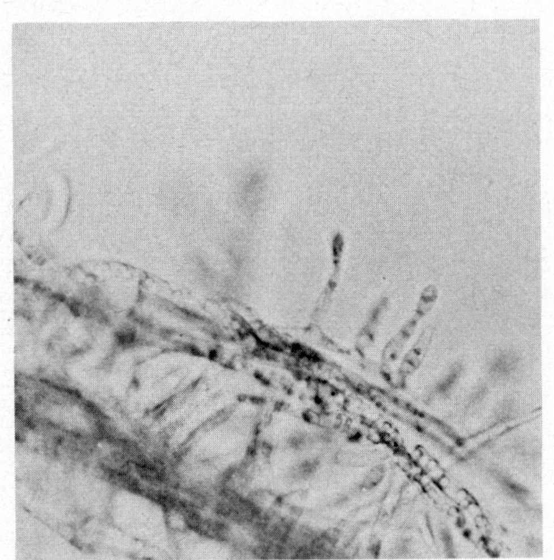

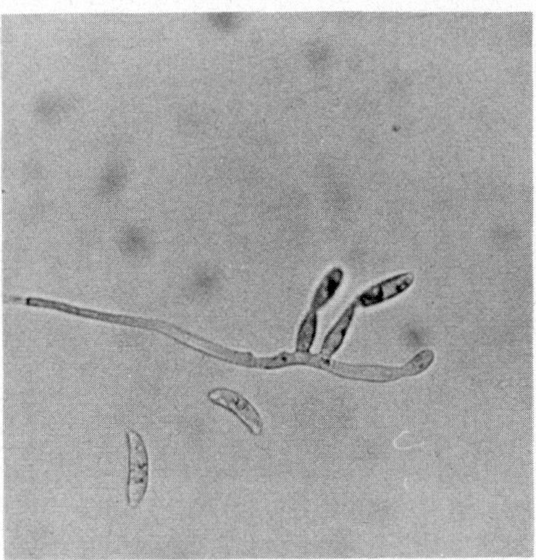

Diagnostic characters : Conidial form and presence of chlamydospores separate it from the related species *F. nivale* and *F. tabacinum*.

(27) FUSARIUM TABACINUM (Beyma) W. Gams, Persoonia 5 : 179, 1968.

Growth rate 3.2 cm.

Cultural pigmentation : colourless to yellowish-salmon or ochraceous becoming light brown.

Conidia 12-16 × 3.4μ, cylindrical, straight or slightly curved with rounded apex and wedge-shaped base.

Chlamydospores not observed.

Perithecial state *Micronectriella cucumeris* (Kleb.) Booth, homothallic ; ascospores 1-septate, 11-15 × 3.4μ.

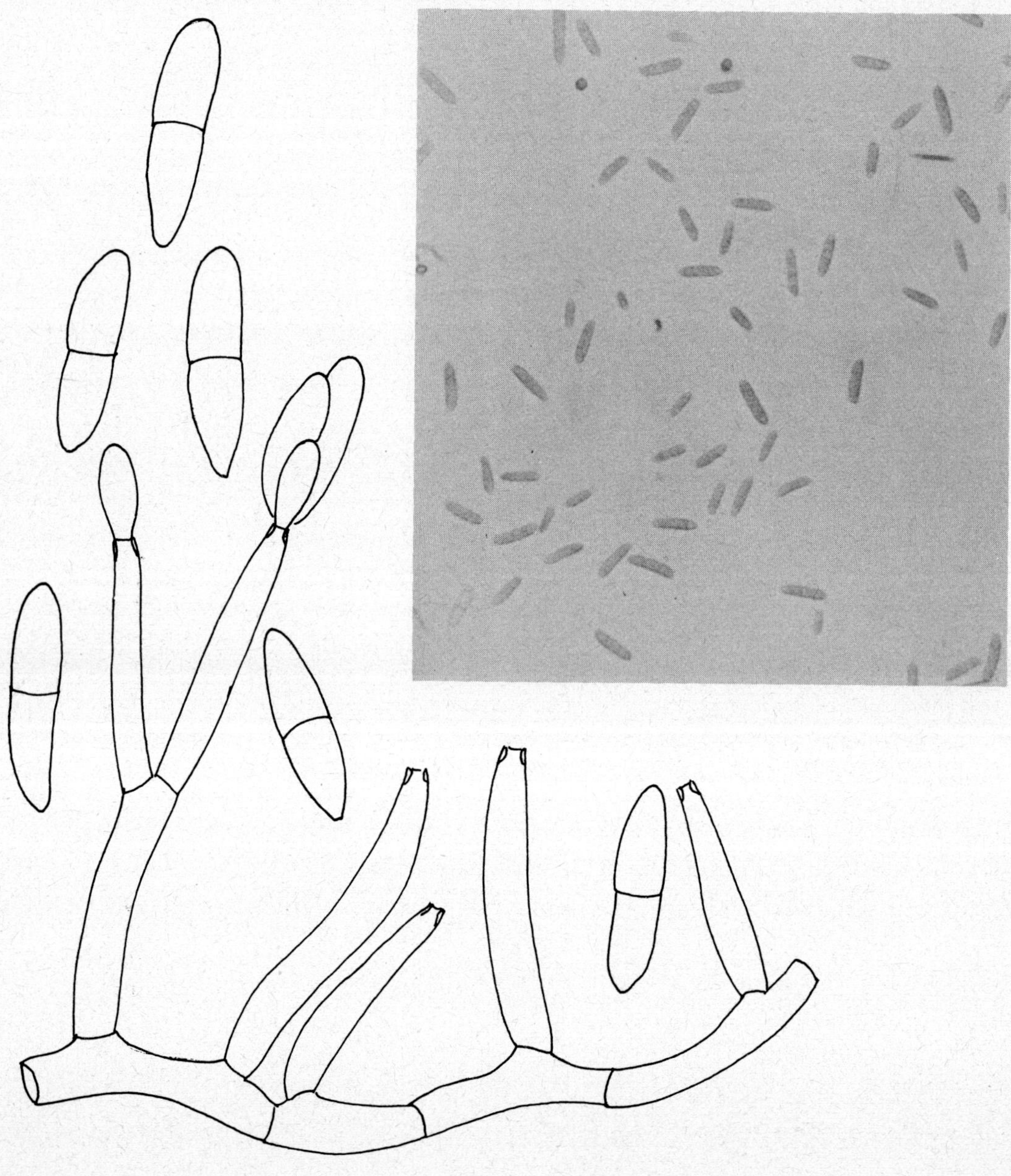

Diagnostic characters : This species is relatively common and represents a marginal or degenerate *Fusarium* species ; aerial mycelium is generally reduced and a pionnotal form of growth is common.

(28) FUSARIUM AQUAEDUCTUUM Lagerh. (Aggr.), *Z. ParasitKde* Abt. 2, **9** : 655, 1891.

Growth rate 0.5 cm.

Cultural pigmentation : pale cream becoming orange or salmon pink with convoluted, merismoid or fibrillose surface appearance.

Macroconidia formed from pionnote sporodochia ; indistinctly 1-5 septate, variable in length, 15-65 × 2.5-4μ.

Chlamydospores absent.

Perithecial state : This is an aggregate species ; some strains produce perithecia referable to *Nectria purtonii* (Grev.) Berk., ascospores 7-11 × 3.5-5μ, but these have not been found in isolates from sewage or polluted water.

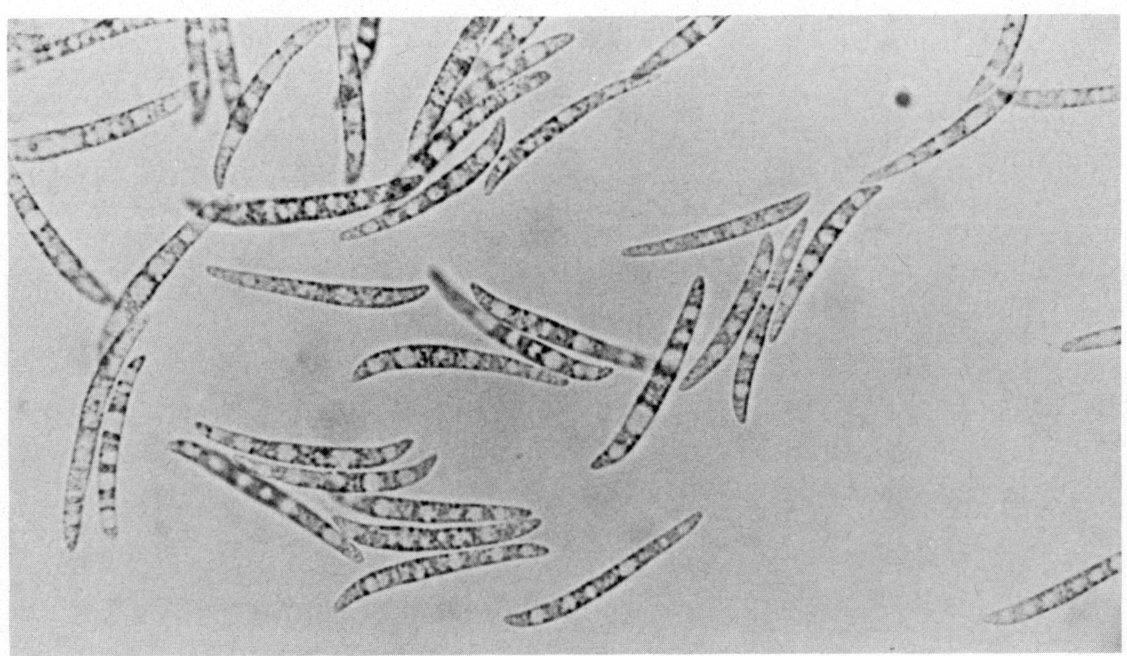

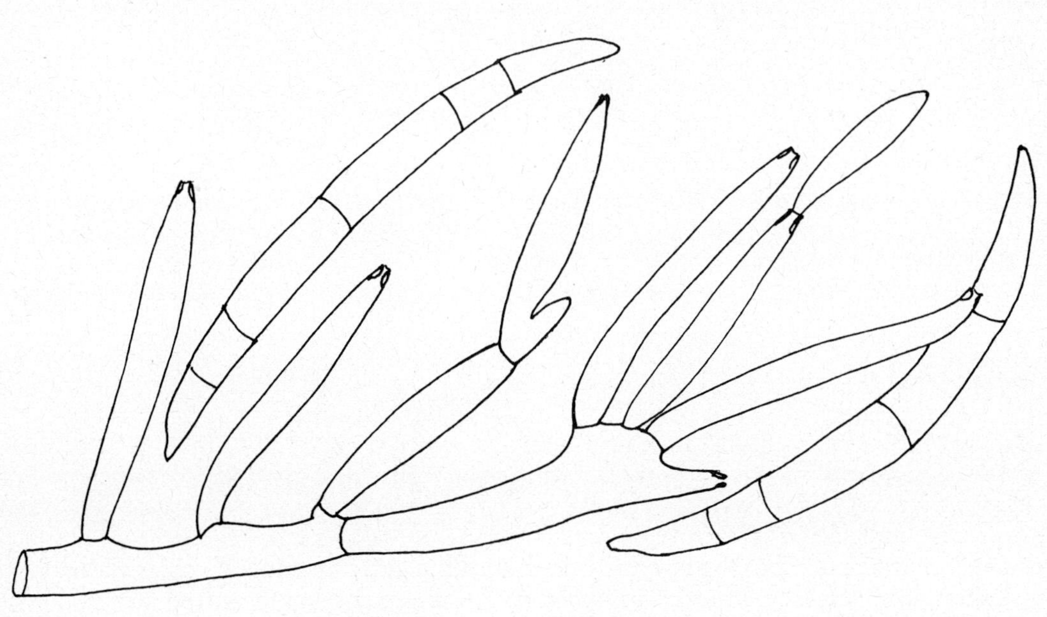

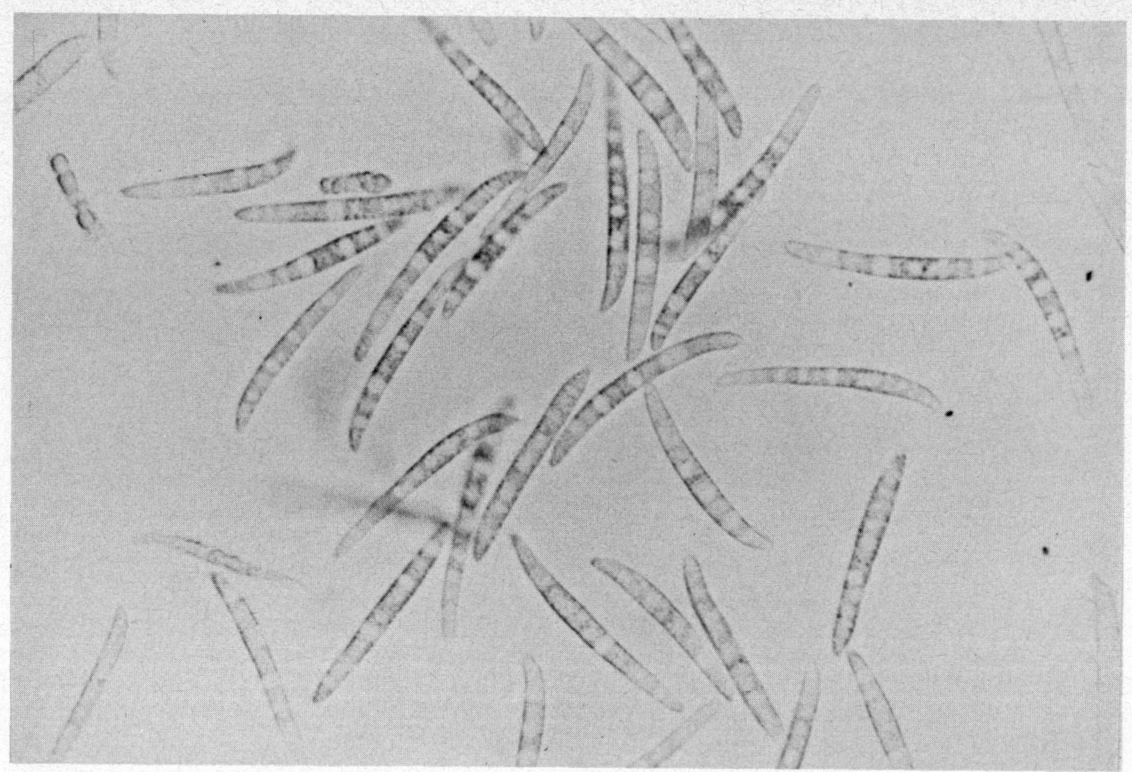

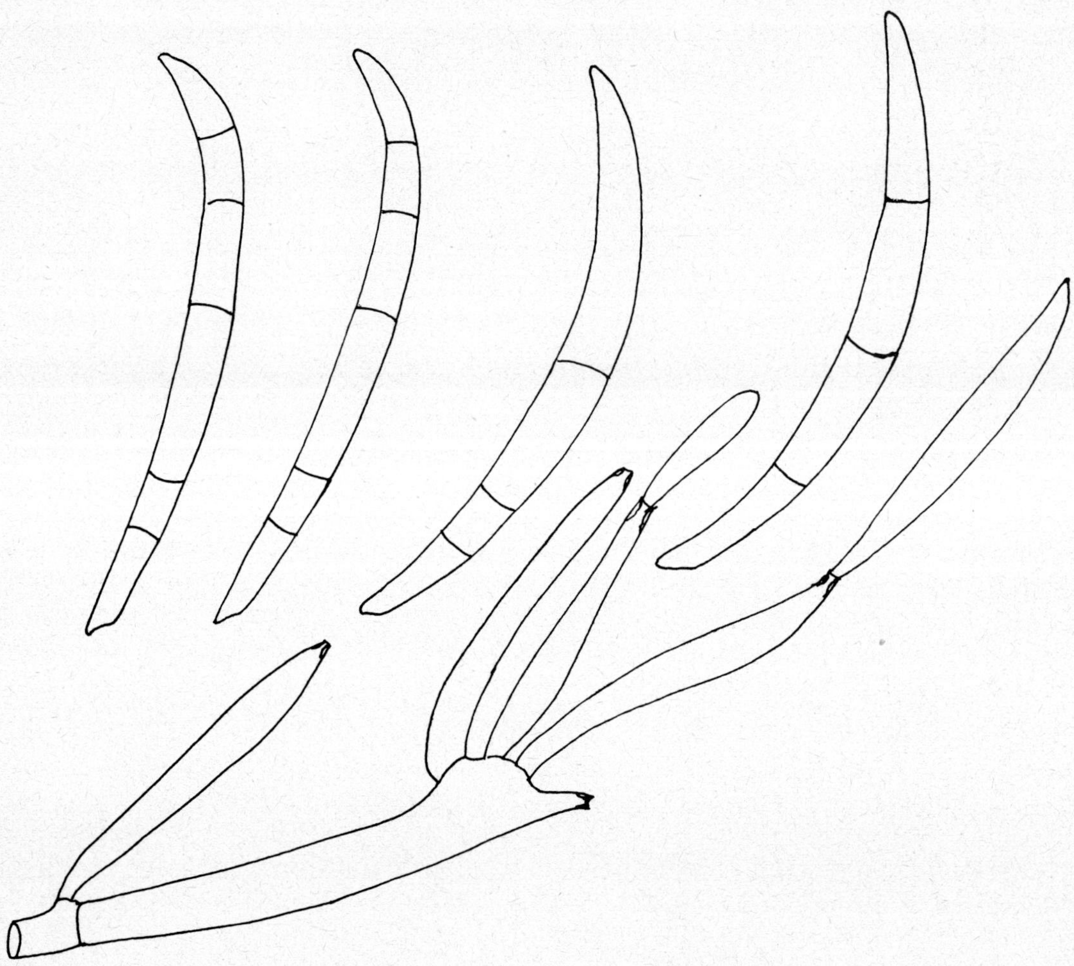

Diagnostic characters : Slow growth, macroconidia and absence of chlamydospores. This description refers primarily to isolates from sewage or polluted water. Other strains with morphologically similar conidia occur as parasites on Sphaeriaceous fungi and may produce perithecia of *Nectria purtonii* (Grev.) Berk.

(29) FUSARIUM MERISMOIDES Corda, *Icones Fungorum* **2** : 4, 1838.

Growth rate 0.9 cm.

Cultural pigmentation : cream, peach to orange.

Macroconidia only are formed, usually from pionnotal sporodochia and cultures
have slimy appearance often without visible mycelium. Macroconidia
initially formed from single or in some strains polyphialides, are 30-70 ×
3.5-4.5μ, aseptate becoming 3-4 septate.

Chlamydospores 8-12μ diam., often slow to form, intercalary, single or in chains.

Perithecial state unknown.

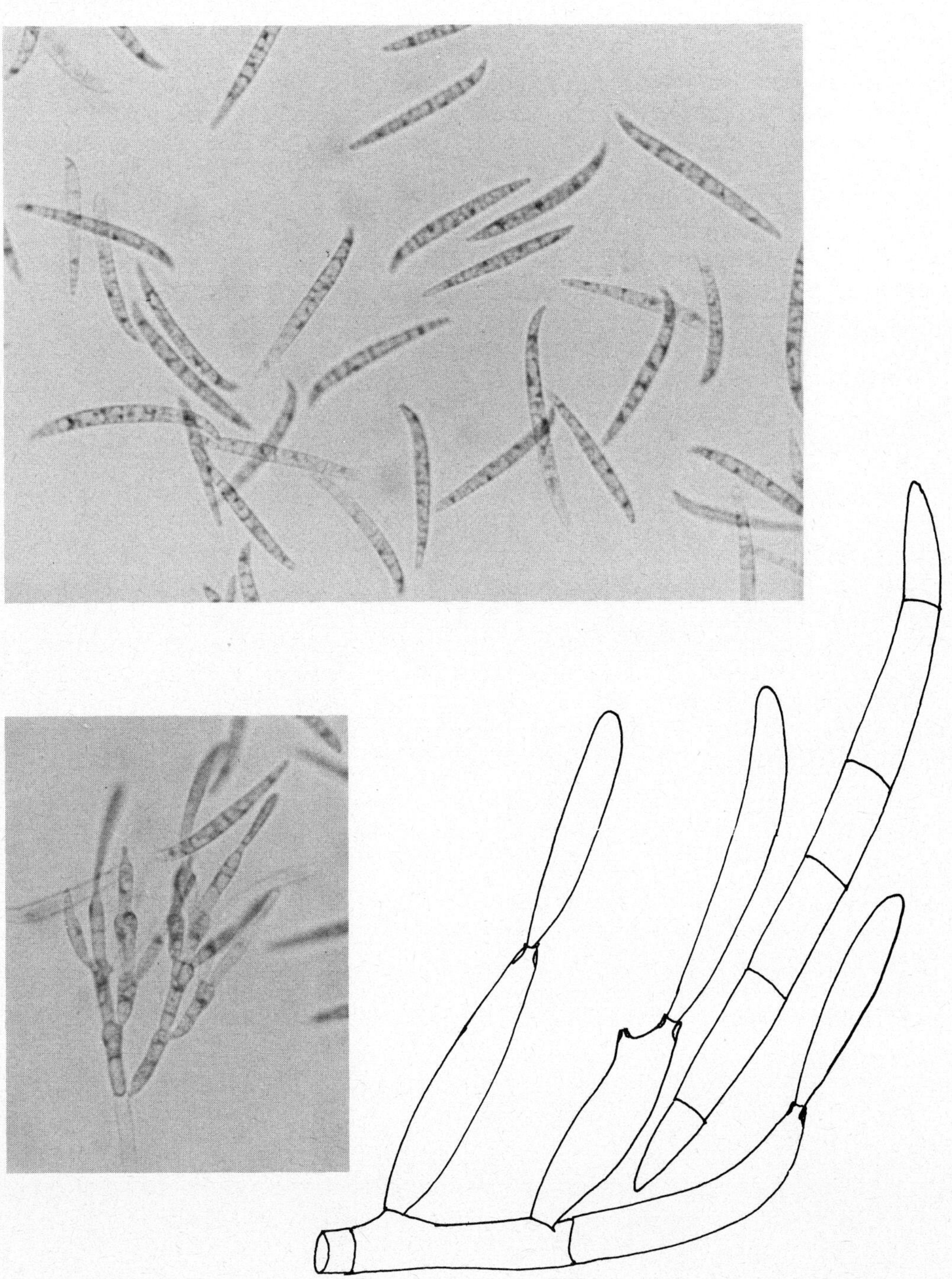

Diagnostic characters : Slow growth rate with slimy pionnotal sporodochia, presence
of polyphialides and spore size.

BIBLIOGRAPHY

Armstrong, G. M.; Armstrong, J. K. (1975) Reflections on the wilt Fusaria. *Annual Review of Phytopathology* **13** : 95-103.

Booth, C. (1971) The genus *Fusarium*. Commonwealth Mycological Institute, Kew, Surrey, England, pp. 1-237.

Cappellini, R. A.; Peterson, J. N. (1965) Macroconidium formation in submerged cultures by a non-sporulating strain of *Gibberella zeae*. *Mycologia* **57** : 962-966.

Gordon, W. L. (1952) The occurrence of *Fusarium* species in Canada. II. Prevalence and taxonomy of *Fusarium* species in cereal seed. *Canadian Journal of Botany* **30** ; 209-251.

Lim, G. (1974) Distribution of *Fusarium* in some British soils. *Mycopathologia et Mycologia Applicata* **52** ; 231-237.

Nash, S. M.; Snyder, W. C. (1962) Quantitative estimations by plate counts of propagules of the bean root rot *Fusarium* in field soils. *Phytopathology* **52** : 567-572.

Nirenberg, H. (1976) Untersuchungen uber die morphologische und biologische differenzierung in der *Fusarium* sektion Liseola. *Mitteilungen aus der Biologischen Bundesanstalt fur Land- und Forstwirtschaft, Berlin-Dahlem,* 169, pp. 1-117.

Papavizas, G. C. (1967) Evaluation of various media and antimicrobial agents for isolation of *Fusarium* from soil. *Phytopathology* **57** : 848-852.

Park D. (1961) Morphogenesis, fungistasis and cultural staling in *Fusarium oxysporum* Snyder & Hansen. *Transactions of the British Mycological Society* **44** : 377-390.

Rayner, R. W. A. (1970) A mycological colour chart. Commonwealth Mycological Institute, Kew, Surrey, England.

Riddell, R. W. (1950) Permanent stained mycological preparations obtained by slide culture. *Mycologia* **42** : 265-270.

Scheffer, R. P.; Walker, J. C. (1953) The physiology of *Fusarium* wilt of tomato. *Phytopathology* **43** : 116-125.

Seemuller, E. (1968) Untersuchungen uber die morphologische und biologische Differenzierung in der *Fusarium*-Sektion Sporotrichiella. *Mitteilungen aus der Biologischen Bundesanstalt fur Land- und Forstwirtschaft, Berlin-Dahlem,* 127, pp. 1-93.

Toussoun, T. A.; Nelson, P. E. (1968) A pictorial guide to the identification of *Fusarium* species. The Pennsylvania State University Press, University Park and London, pp. 1-51.

Toussoun, T. A.; Nelson, P. E. (1975) Variations and speciation in the Fusaria. *Annual Review of Phytopathology* **13** : 71-82.